01-CTS-22

RESEARCH PRIORITIES FOR SUCCESSFUL ASSET MANAGEMENT: A WORKSHOP

2002

Water Environment
Research Foundation

The Water Environment Research Foundation, a not-for profit organization, funds and manages water quality research for its subscribers through a diverse public-private partnership between municipal utilities, corporations, academia, industry, and the federal government. WERF subscribers include municipal and regional water and wastewater utilities, industrial corporations, environmental engineering firms, and others that share a commitment to cost-effective water quality solutions. WERF is dedicated to advancing science and technology addressing water quality issues as they impact water resources, the atmosphere, the lands, and quality of life.

For more information, contact:
Water Environment Research Foundation
601 Wythe Street
Alexandria, VA 22314-1994
Tel: (703) 684-2470
Fax: (703) 299-0742
www.werf.org
werf@werf.org

For non-subscriber sales information, contact:

IWA Publishing
Alliance House, 12 Caxton Street
London SW1H 0QS, United Kingdom
Tel: +44 (0) 20 7654 5500
Fax: +44 (0) 20 7654 5555
www.iwapublishing.com
publications@iwap.co.uk

Library of Congress Catalog Card Number: 2002115043
Printed in the United States of America
WERF ISBN: 1-893664-66-X IWAP ISBN: 1-84339-659-9

CH2M HILL

The research on which this report is based was funded, in part, by the United States Environmental Protection Agency through Cooperative Agreement No. CP-829433-01-0 with the Water Environment Research Foundation (WERF). Unless an EPA logo appears on the cover, this report is a publication of WERF, not EPA.

Acknowledgments

This report was produced and published by the Water Environment Research Foundation to summarize the findings of a workshop held on March 21-22, 2002 in Alexandria, Va. The workshop addressed asset management programs across public water and wastewater utilities and recommended a research agenda to promote the next generation of tools for reducing risk and improving competitiveness.

We wish to thank the workshop participants and the Project Subcommittee for their active participation in the workshop and contributions to this document. In addition, special thanks are extended to Steve Allbee and Charles Vanderlyn for their contribution and assistance and Marsha Passoja for her assistance with the logistics of the workshop.

Workshop Participants

Alan P. Landis .... Metropolitan Sewer District of Greater Cincinnati, Ohio
Andrew Speers .... Commonwealth Scientific and Industrial Research Organization, Australia
Anthony Wilson .... New Plymouth District Council, New Zealand
Art Hamid .... Montgomery Watson Harza, Inc., Georgia
Betty Matthews-Malone .... City of Hamilton, Ontario
Charles Vanderlyn .... U.S. Environmental Protection Agency, OWM
Cuthbert Braveboy .... District of Columbia Water and Sewer Authority
David Binning .... Fairfax County Water Authority, Virginia
Douglas M. Stewart .... Orange County Sanitation District, California
Duncan Rose .... Parsons, Florida
Eric de Rooy .... Sydney Water Corporation, Australia
Jim Patterson .... Columbus Water Works, Georgia
Kenneth Harlow .... Brown and Caldwell, California
Kevin Young .... Hunter Water Corporation, Australia
Mick Bourke .... City West Water Ltd., Australia
Paul Causey .... Delta Diablo Sanitation District, California
Ralph L. Charlton .... Alexandria Sanitation Authority, Virginia
Ray Yep .... Santa Clara Valley Water District, California
Robert C. Weaver .... Kelly & Weaver, Washington DC
Roger J. Dolan .... Dolan Environmental Services, California
Ron Chapman .... Advantica Technology, United Kingdom
Russell Shaw .... UMS Group, Australia
Stephen T. Hayashi .... Union Sanitary District, California
Stewart Ramsey .... UMS Group North America Inc., New Jersey
Terrance M. Brueck .... EMA, Inc., Minnesota
Warwick Busch .... Meritec Ltd., New Zealand
Wayne Green .... City of Toronto, Ontario

Observers

Anthony N. Tafuri .... U.S. Environmental Protection Agency, ORD
Barry Benroth .... U.S. Environmental Protection Agency, OWM
Bijlsma Luitzen .... World Bank - Netherlands Water Partnership Program, Washington, DC
Doug Abbott .... Maryland Center for Environmental Training, College of So. Maryland
Robert S. Morrison .... Jason Consultant International, Washington DC
Timothy J. Dwyer .... U.S. Environmental Protection Agency, OWM
Walton Brodtman .... U.S. Environmental Protection Agency, OWECA

Project Subcommittee

Daniel K. O'Brien .... Massachusetts Water Resources Authority, Massachusetts
Michael Sweeney, Ph.D., P.E. ... Louisville & Jefferson County Metropolitan Sewer District, Kentucky
Scott Haskins .... Seattle Public Utilities, Washington
Steve Allbee .... U.S. Environmental Protection Agency, OWM
Tyler Richards, PSC Chair .... Gwinnett County, Georgia

Project Investigators (PI) Team

Edward Wagner .... CH2M HILL
Elisa Speranza .... CH2M HILL
John Rogers .... CH2M HILL
Lisa Bacon .... CH2M HILL
Alan Ispass .... PA Consulting Group
Cara Bobchek .... Capstone Communications

WERF Team

Linda Blankenship .... Deputy Director, Development and Subscriber Services
Amit Pramanik .... Director, Technical Services
Roy Ramani .... Project Manager

Table of Contents

Executive Summary

Project Goals

This report details the presentations, discussions, workgroup activities, and workshop consensus on recommended research projects, and contains a conceptual budget and scope for the proposed research projects.

Developing an asset management research agenda will provide Water Environment Research Foundation subscribers with improved tools and approaches to improve their competitiveness, reduce risk, and achieve environmental and customer service goals. Public water and wastewater utilities have a unique opportunity to benefit from developing and implementing an asset management program. Asset management research will be applicable to all infrastructure assets managed by public water, wastewater, stormwater and drainage utilities.

"Asset management will become our principal business in the water and wastewater industry." -Workshop Participant

WERF convened a workshop on March 21-22, 2002, for the purpose of characterizing the spectrum of asset management programs across public water and wastewater utilities and recommending a research agenda to promote the next generation of tools for reducing risk and improving competitiveness. WERF's goals for the workshop were to:

*For utilities, asset management will involve **change management**, affecting everything from the pricing of services to the way the utility is organized to deliver those services, as well as benchmarking and information exchange among agencies.*

- Survey, examine, and document effective asset management approaches
- Look at the state of implementation of asset management and identify barriers to more broad-based implementation
- Pinpoint where improved tools and technical approaches could enhance decision making, confidence, efficiency, communication, and results
- Determine differences between large and small utilities' approaches to asset management
- Prepare a prioritized list of research initiatives to address the industry needs and develop a conceptual scope and budget for these research projects

A recent WERF Subscriber Survey supports the view that asset management is a priority issue for utilities. Utility respondents ranked asset management fifth among 20 categories by priority, and sixty percent of respondents ranked asset management a first or second priority.

"Because of the dialogue we are having here, this workshop presents an opportunity for us to become leaders in the field of asset management". –Workshop Participant

"We know there are different definitions for the term 'asset management' around the world. We need to define the term in a way that everyone can understand." -Workshop Participant

Profile of the Workshop Participants

WERF invited utility managers, regulatory officials, and consultants from throughout the United States and Canada. Criteria for selecting participants included factors such as the size, location, type, and age of their utility. In addition, WERF invited representatives from both the utility and consulting fields from Australia, New Zealand, and the United Kingdom, all countries recognized to have advanced asset management experience.

Workshop Approach

The WERF workshop was facilitated by a team of CH2M HILL and PA Consulting Group, who also developed the Association of Metropolitan Sewerage Agencies' Managing Public Infrastructure Assets, published in 2002 in cooperation with the American Water Works Association, the Water Environment Federation, and the Association of Metropolitan Water Agencies.

A team led by experts in asset management facilitated the workshop. Through presentations, plenary sessions, and "breakout" workgroups, the facilitation project team led the workshop participants through the process of developing criteria for asset management research needs, identifying barriers, and assessing gaps in programs, tools, and techniques. Using decision science techniques, the participants then developed a list of new research initiatives, to which they assigned priorities, conceptual scopes, and budgets.

Research topics recommended through the workshop fall into the following categories:

- Strategic Asset Management
- Knowing What You Have
- Maintaining Your Assets
- Renewal and Replacement
- Meeting Future Needs
- GASB 34
- Integrated Asset Management Systems
- Business Process Redesign
- Outreach and Reporting

The 11 projects ultimately recommended as a result of the workshop represent combinations of 18 of the 28 projects originally suggested by the workshop participants. Several suggested by workshop participants were not recommended in the final group or were excluded because they were considered to be premature or they do not meet WERF's criteria or the profile of the kind of research that WERF typically funds.

The 11 recommended research projects, along with an estimated duration and budget for each, are summarized in order of priority in the table below.

Recommended Projects

Priority/ Project No. *	Project Description	Duration	Cost (000s)
1 (W-16, W-17, W-20, W-26)	Develop protocols for assessing the condition and performance of water and wastewater infrastructure assets and develop predictive models for correlating asset condition and performance	Phase 1: 12 mos. Phase 2: 12 mos. Phase 3: 12 mos.	Phase 1: $200 Phase 2: $200 Phase 3: $250
2 (W-4, W-24, W-27, W-26)	Construct predictive life-cycle models for water and wastewater infrastructure that project life-cycle costs and risks	Phase 1: 12 mos. Phase 2: 24 mos. Phase 3: 12 mos.	Phase 1: $150 Phase 2: $250 Phase 3: $200
3 (W-9)	Construct a template for preparing infrastructure asset management plans	6 mos.	$75
4 (W-3)	Prepare guidance for asset management strategic planning	18 mos.	$200
5 (W-25)	Develop practical methodologies for calculating life-cycle costs for water and wastewater infrastructure assets	18 mos.	$250
6 (W-8)	Document case studies of improved customer service and satisfaction and reduced costs as a result of implementing asset management	18 mos.	$200
7 (W-21)	Identify O&M best practices by asset category, condition and performance requirements	24 mos.	$250
8 (W-18)	Define best practices for the integration of water and wastewater databases	18 mos.	$250
9 (W-6, W-14, W-19)	Establish methodologies for determining water and wastewater asset value, compiling asset inventories and capturing and compiling asset attribute information	Phase 1: 12 mos. Phase 2: 12 mos. Phase 3. 12 mos.	Phase 1: $100 Phase 2: $100 Phase 3. $150
10 (W-12)	Develop a framework for asset management accountability	18 mos.	$150
11 (W-15)	Assess the feasibility of establishing an Asset Management Standards Board for the water and wastewater industry	12 mos.	$50

*The table in the Appendix lists each of the 28 research projects as originally suggested by the workshop participants, cross-referenced by project number (beginning with "W," to indicate that the project is presented as suggested at the workshop) and priority number recommended by the workshop project team (1-11).

Chapter 1: Premises and Foundations of the Workshop

Promoting Asset Management

"Asset management will become our primary business."-Workshop Participant

Wastewater and water utilities are challenged by increasing costs and escalating demands from customers and regulatory agencies to reduce risks and improve service. In addition, there is a growing realization that utility assets such as pipelines, no matter how long-lived, must be proactively managed. In the United States, asset management concepts and processes are rapidly evolving, and many utilities are actively pursuing their implementation.

For water and wastewater utilities, asset management will involve ***change management****, affecting things such as the pricing of services, how a utility is organized to deliver the services, information exchange among agencies, and benchmarks.*

Although a number of techniques and tools exist to support and effectively implement asset management programs, utilities are struggling to identify, evaluate, and select those most appropriate for their specific needs. Furthermore, all public utilities, large and small, need additional techniques and tools to implement programs that will efficiently achieve the full benefits of comprehensive asset management.

Public wastewater and water utilities have a unique opportunity to benefit from implementing an asset management program. Asset management ensures that the best decisions are made throughout asset life cycles to reduce risks, optimize performance, and minimize costs. The challenge a utility faces in developing an asset management program is to adopt appropriate business processes, institute strategic management approaches, and install the necessary analytical and decision-making tools so as to achieve the greatest benefits for its stakeholders.

Workshop Goals

The Water Environment Research Foundation convened a workshop on March 21-22, 2002, to characterize the spectrum of asset management programs across public water and wastewater utilities and recommend a research agenda to promote the next generation of tools for reducing risk and improving competitiveness. Developing an asset management research agenda will provide WERF subscribers with improved tools and approaches to increase competitiveness, reduce risk, and achieve environmental and customer service goals.

WERF's goals for the workshop included the following:

- Survey, examine, and document effective asset management approaches
- Look at the state of implementation of asset management and identify barriers to broader implementation
- Pinpoint where improved tools and technical approaches could enhance decision-making, confidence, efficiency, communication, and results
- Determine and distinguish differences between large and small utilities' approaches to asset management
- Prepare a prioritized list of research initiatives to address the industry needs and develop a conceptual scope and budget for these research projects

The workshop agenda is shown in Appendix A.

Workshop Participants

WERF invited utility managers, regulatory officials, and consultants from throughout the United States and Canada to the workshop. Criteria for selecting participants included factors such as their utilities' size, location, type, and age, with the goal of achieving a balance of key utility and management characteristics:

- Water and wastewater
- Funding sources (general purpose funds, user charge, full enterprise funding
- Size
- Geographic location
- Age of system
- Municipal agency vs. special purpose district
- Single jurisdiction vs. wholesale services to multiple political jurisdictions; serve satellite systems
- Agencies with dedicated renewal and replacement funds vs. all capital work considered as individual projects competing for funding
- Presence of conveyance (distribution/collection) in addition to treatment plants
- Have begun to implement comprehensive/integrated asset management programs
- Participants' functional responsibilities (balance of planning, operations, engineering, finance, executive)

According to a representative of WERF's Project Subcommittee:

- *GASB standards will cause utilities to account for their assets differently than they have in the past*
- *Privatization and regulatory issues are affecting how we care for assets*
- *Defining "asset management" is a goal of the Project Subcommittee: the term is defined differently across the world*

In addition, WERF invited representatives from both the utility and consulting fields from Australia, New Zealand, and the United Kingdom, all countries recognized to have advanced asset management experience. A complete list of workshop attendees is presented in Appendix B.

Workshop Facilitation Project Team

The consultant team of CH2M HILL and PA Consulting Group facilitated the workshop. This team previously researched and wrote the Association of Metropolitan Sewerage Agencies (AMSA) 2001 handbook, *Managing Public Infrastructure Assets to Minimize Cost and Maximize Performance* and developed regional asset management workshops for utility managers. Through the development of the handbook and the regional workshops, this team investigated implementation of asset management nationally and internationally. The team includes three active and well-known former utility managers. These industry leaders understand the technical, political, fiscal, customer, and management demands on public utilities.

The workshop project team assisted WERF by:

- Identifying potential workshop participants
- Researching and assessing benchmark asset management concepts and processes—in other words, the state of the practice. The AMSA-commissioned handbook, *Managing Public Infrastructure Assets*, was the primary reference for this research
- Researching the types of techniques and tools required for supporting and conducting asset management programs, including those demonstrated to be effective outside the wastewater and water industry and outside the United States, specifically in New Zealand and Australia
- Conducting a pre-workshop survey of participants to learn about the status and level of asset management activities in their utilities. (See Appendix E for sample survey form)

This report details the presentations, discussions, workgroup activities, and workshop consensus on recommended research projects. Based on the workshop activities, WERF's Project Subcommittee and project team have developed the comprehensive prioritized research program presented here for creating the next generation of tools to propel wastewater and water utilities toward achieving the full benefits of asset management—that is, to reduce risk, improve competitiveness, and enhance customer service.

Drivers for Asset Management

Customer service, competitiveness, GASB Statement No. 34 and EPA's draft CMOM rules, as well as security issues, represent major drivers toward asset stewardship. It is not a question of whether asset management will become a new core competency; it is a question of when.

New business models as well as regulatory and security issues represent both an opportunity and a challenge in asset management for utilities, states, and local governments. Utility managers will have the opportunity to more accurately report the condition and performance of their infrastructure assets. They also will be challenged to develop practical, affordable, and focused methods and tools that help asset mangers get the most performance from their assets over the life of the asset for the dollar invested. The AMSA handbook *Managing Public Infrastructure Assets to Minimize Costs and Maximize Performance* provides a good basis of best practices to begin to apply toward achieving these goals.

Asset management requires excellent cross-functional relationships within a utility. In particular, engineers and operators must work more closely with financial managers.

Asset management approaches should focus on the payback of investments, as well as the costs, and take a comprehensive view of all infrastructure resources, their use, impact on the utility's mission, and their value to customer service. U.S. Environmental Protection Agency studies have shown that asset management can substantially reduce operations and maintenance costs. In a recent Water Environment Research Foundation Workshop, our colleagues at Hunter Water in Australia reported that they reduced their O&M costs by 50 percent using their comprehensive asset management approach that is focused on risk-based decisions.

The basic elements of a good asset management program—strategy, asset retention, tool integration, business process redesign, and outreach and reporting--are shown in the triangle diagram below:

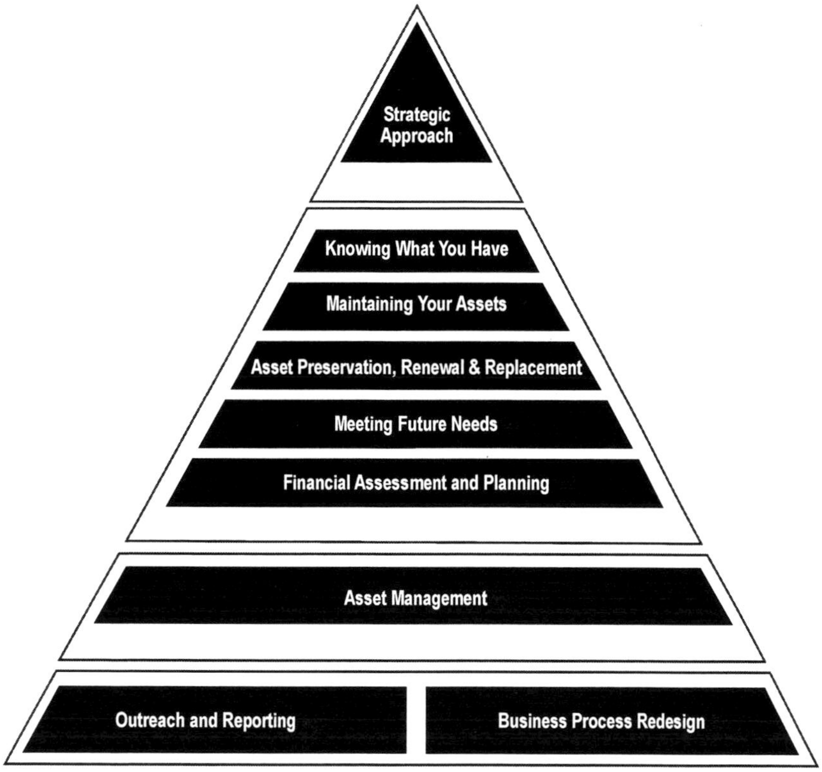

WERF Subscriber Survey

Periodically, WERF surveys its subscribers regarding areas of interest. Charles Noss of WERF presented preliminary highlights from the most recent survey concerning asset management:

- Respondents ranked 20 categories by priority
- Of 90 respondents, 60 represented utilities
- Among all 90 respondents, asset management was ranked seventh in priority among 20 categories
- Of the 60 utility respondents, asset management was ranked fifth out of 20 categories
- Sixty percent of respondents ranked asset management first or second in priority
- Utility respondents ranked four items higher than asset management: collection systems, TMDLs, odors, and biosolids

These preliminary results indicate that asset management is a topic of interest to WERF subscribers and highlight the timeliness of the asset management workshop.

Chapter 2: State of the Practice

"Asset management is a means of managing infrastructure to minimize the cost of owning and operating it while delivering the service levels that customers desire." Managing Public Infrastructure Assets, AMSA, 2002.

The Current State of Asset Management

The workshop agenda included a presentation on the state of the practice of asset management in the water and wastewater industries. This set the baseline for exploration of moving to a more objective process for asset management, defending the decisions made during the process, and effectively communicating the decisions to stakeholders. Highlights of this presentation are provided below.

Strategic Asset Management

Aligning capital investments and infrastructure maintenance with the overall mission helps achieve a utility's goals:

- Clarify agency mission and adopt asset management objectives that guide future investment decisions
- Charter a team to establish leadership and cross-functional communication
- Prepare an asset management strategy to ensure a successful program
- Move toward life-cycle decision making to find long-term, lower-cost solutions

The utility's Asset Management Team develops the fundamental strategic elements of the plan:

- New and revised policies
- Role of each functional unit toward asset management
- Written procedures for internal and external communication
- Service and performance levels related to asset function and condition
- Implementation program including improved decision and data management processes and tools

Asset management involves all phases of an asset's life cycle:

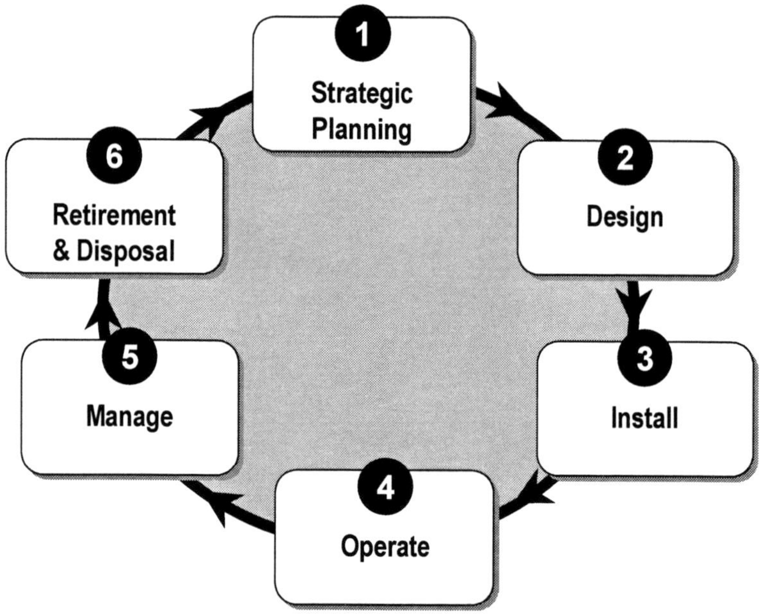

Organizations benefit from asset management in the following areas:

Decision making using life-cycle analysis ensures lowest cost solutions:

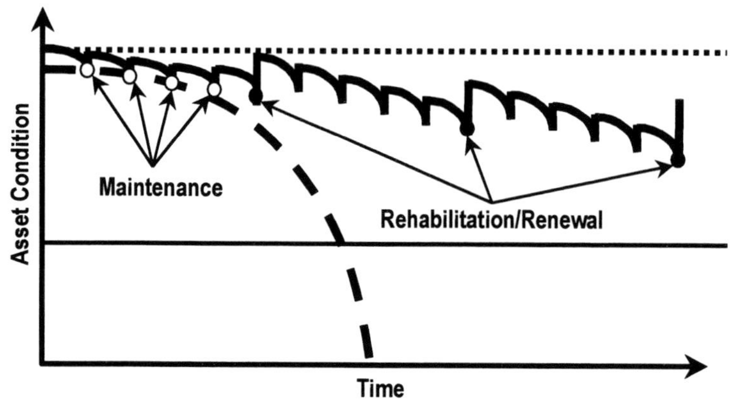

- ***Rates:** Minimize the cost of service by constructing system improvements and repairs with the lowest life-cycle ownership costs that meet service-level objectives.*

- ***Maintenance and Replacement:** Replace the current high percentage of reactive maintenance with planned maintenance, and replace or repair assets based on known asset condition and forecasts of condition indexes and risk indexes. This ensures a solid basis for risk management and that those investments are neither too early nor too late.*

Asset management establishes the platform for the total life-cycle approach, which:

- Includes all assets to varying levels of detail
- Tracks costs and condition
- Relates condition to performance and risk
- Projects condition, performance, risk, and cost
- Evaluates alternatives to determine appropriate investment decisions
- Considers real world constraints
    - Service levels
    - Revenue availability
    - Compliance
    - Political factors

- ***Funding:*** *Compile information in flexible graphic formats to show Boards or other funding authorities the forecasts for the financial advantages of each project and the financial implications of delaying needed repairs or replacements.*
- ***Planning:*** *Proactively Conduct all the above activities within a structured framework to reduce risks.*

Knowing What You Have

An accurate asset inventory sets the stage for effective management:

- Bring together information from various departments into a core database to facilitate long-term management
- Use cost and risk factors to identify what level of asset tracking detail will streamline the inventory report
- Reflect total costs in the asset value for developing accurate capital renewal and replacement strategies
- Develop consistent, replicable condition grading systems linked to performance and objectives

Reconciling information located in various departments into a core asset inventory database facilitates long-term planning:

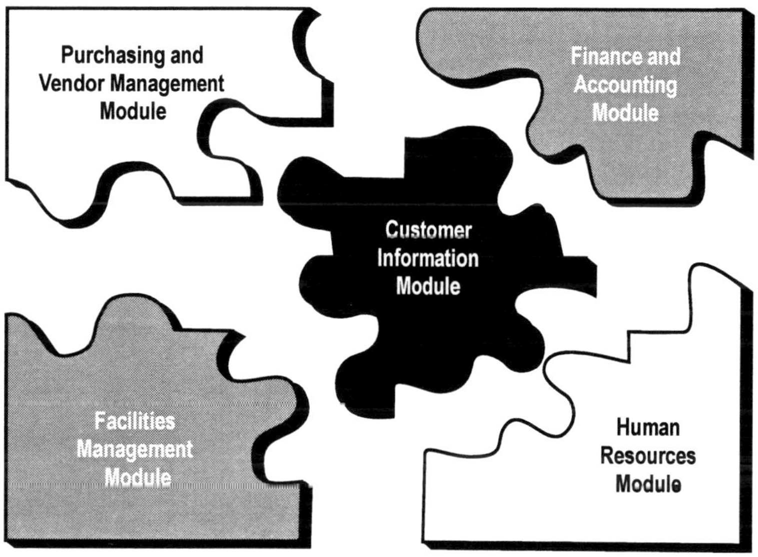

"Customers' demands will change. The utility manager must figure the true cost of higher levels of service—not just the current level. We should be able to show customers the range of service levels available at their corresponding costs, so that they will understand what level of service they can expect for what price." -Workshop Participant

Using cost and risk factors helps to identify levels of detail necessary for tracking assets:

- Require a significant investment to repair or replace?
- Impact the utility's goals?
- Create a health or safety issue?
- Cause an environmental impact?
- Result in a permit violation or other regulatory/legal concern?
- Create negative publicity or undermine public confidence?

A consistent grading scale creates a common vocabulary, helps illustrate risk, and prioritizes operations and investments:

Grade	Condition	Description
0	Abandoned	Asset abandoned (no longer in use) or no longer exists
1	Very Good	Sound physical condition. Meets current needs. Operable and well-maintained. Asset expected to perform adequately with routine maintenance for 10 years or more. No work required.
2	Good	Acceptable physical condition. Shows minor wear that has minimal impact on performance. Minimal short-term failure risk. Potential for deterioration or impaired performance over next 5-10 years. Minor work (if any) required.
3	Fair	Functionally sound but showing wear and diminished performance. Moderate short-term failure risk. Potential for further deterioration and diminished performance within next 5 years. Renewal or major component replacement expected within next 5 years. Minor work required but asset is serviceable.
4	Poor	Asset functions but requires high level of maintenance to remain operational. High risk of short-term failure. Likely to have significant deterioration in performance within next 2 years. Renewal or replacement expected within next 2 years. Substantial work required, asset barely serviceable.
5	Very Poor	Asset failed or failure is imminent. Excessive maintenance required. No further service life expectancy. Significant health and safety hazard. Major work or replacement is urgent.

Maintaining Your Assets

Targeted use of labor, equipment, and materials prevents service interruptions and saves money:

- Each approach is a legitimate component of a maintenance strategy
  - Planned maintenance (preventive, predictive)
  - Unplanned maintenance (reactive)

Balanced maintenance achieves an appropriate level of planned work to obtain cost effective performance:

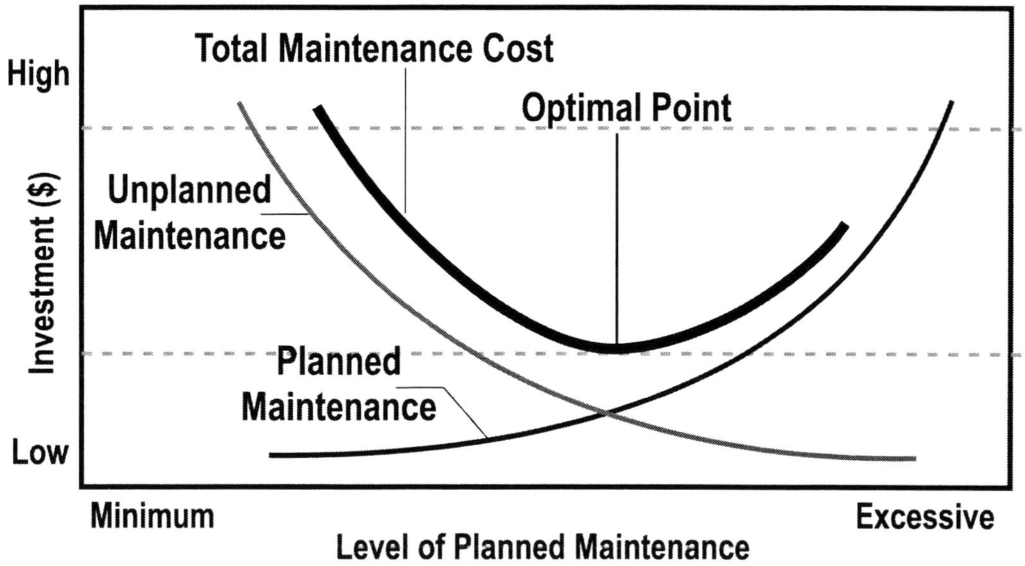

A maintenance management program provides opportunities to track costs, condition, and performance.

Asset Preservation, Renewal, and Replacement

Planned renewal and replacement preserve an asset's performance and replace it at the right time:

"We let our buried assets run to failure 90 percent of the time, based on our understanding of our customers' desires." -Workshop Participant

- Establish a renewal and replacement program that provides long-term perspective
- Define current problems and anticipate future needs so as to reduce surprises and improve reliability
- Analyze management alternatives for life-cycle decision-making to minimize long-term costs
- Assess deficiencies and communicate how they create risks for your utility

Repair and replacement almost always results in lower life-cycle costs than "run to failure":

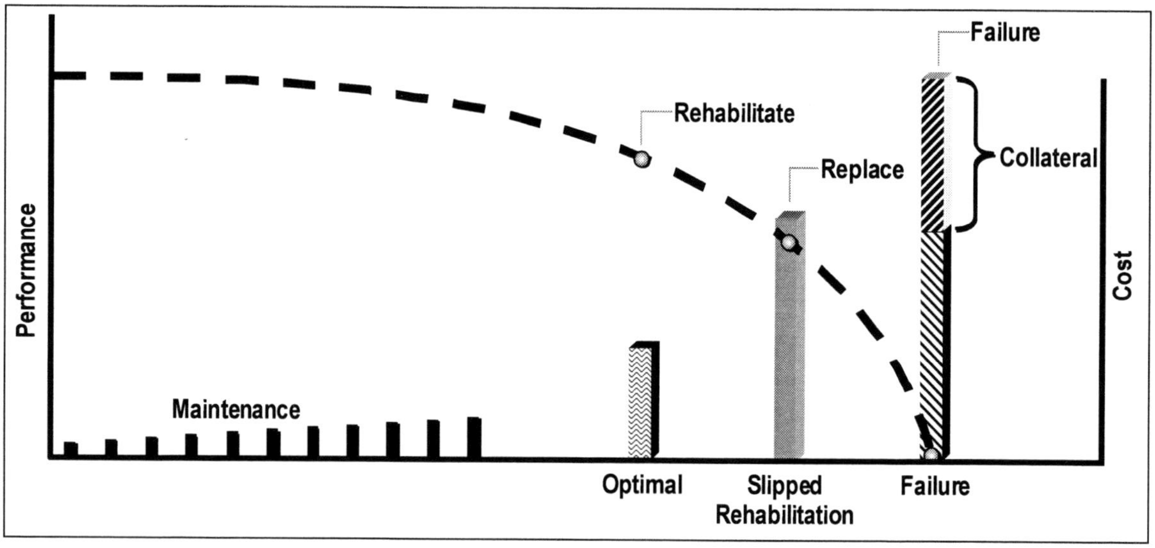

Capital renewal schedules provide reasonable projections of long-term needs:

- Schedule for the life of the asset, including its replacement

"We aren't convinced that there's a statistical model that works to predict the reliability of assets." - Workshop Participant

Predictive modeling examines all factors impacting long-term performance to identify need for repair and replacement:

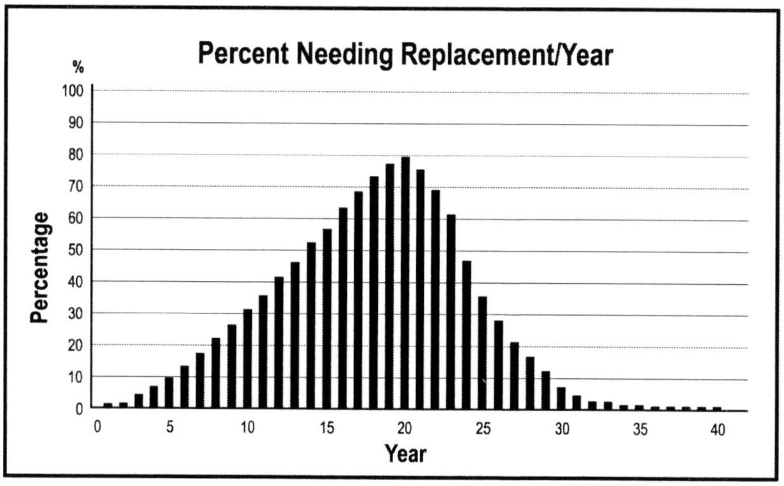

Risk assessment plus life-cycle cost projections allow for informed decisions:

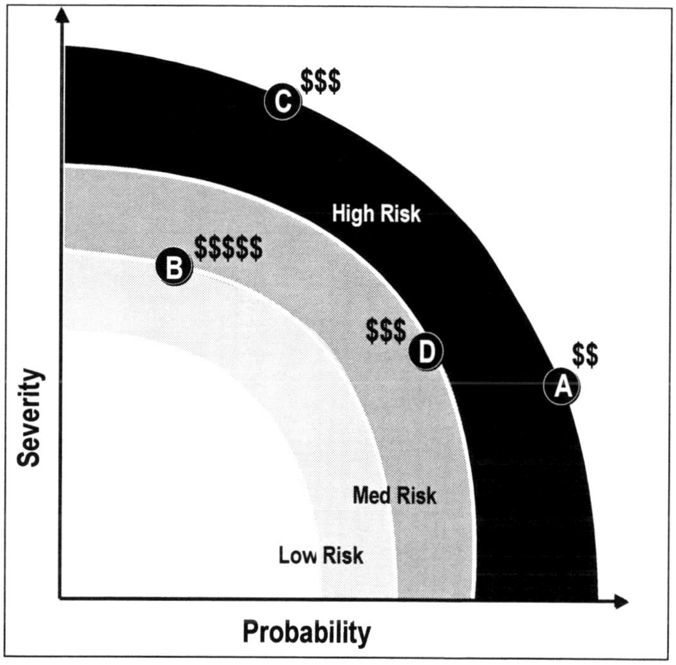

Meeting Future Needs

Balancing system investment requirements is necessary to meet future capital needs:

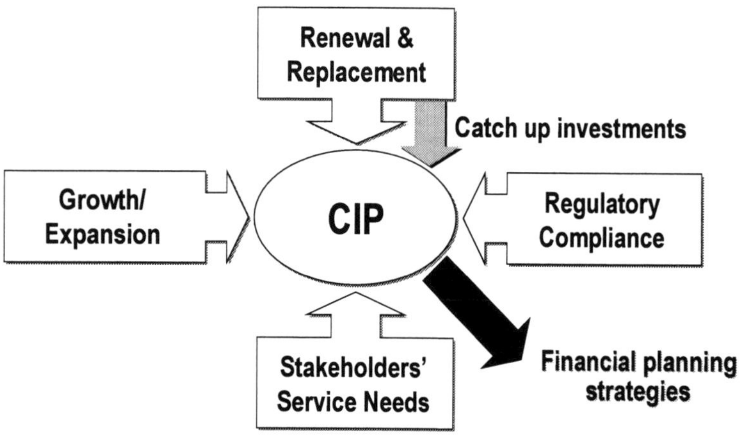

Developing a long-term capital planning horizon defines total long-term funding needs and can balance revenue plans:

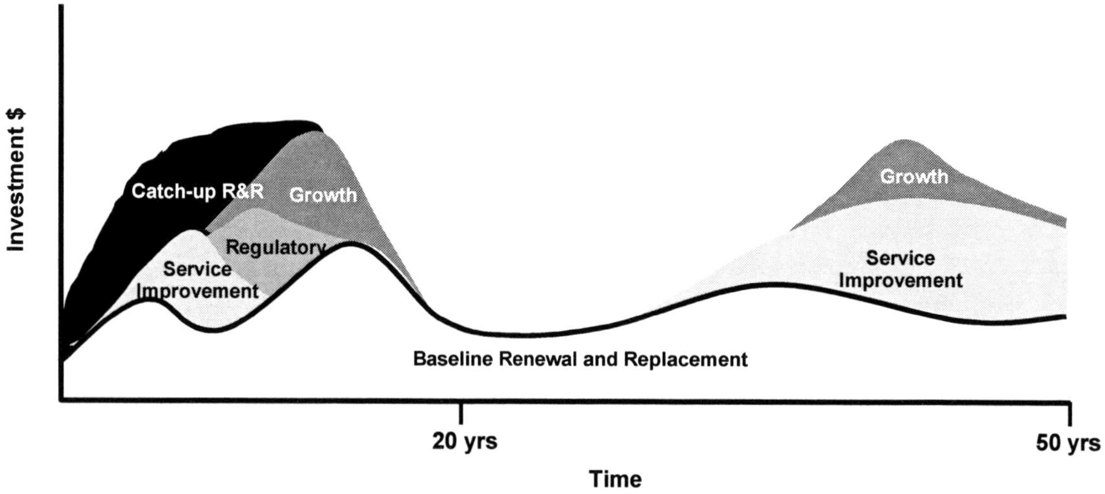

"An important driver for asset management is having a real understanding of what your customers want. How much failure will customers accept? What service level must you maintain to satisfy them?" -Workshop Participant

Translating revenue needs to impact individual customers demonstrates program feasibility:

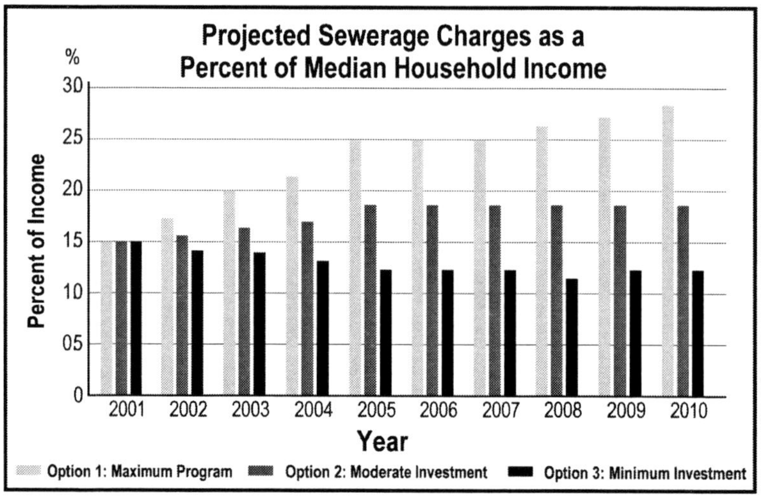

Complying with GASB 34

"Finance people aren't concerned with asset condition. They won't be the ones driving us to asset management." -Workshop Participant

"Utility managers have to decide to have a better command of the future of their system and pull the accounting community along with them—not vice versa." -Workshop Participant

Government Accounting Standards Board Statement 34 (GASB 34) reporting requirements signal future asset management expectations:

- Depreciation
    - Complete, up-to-date inventory
    - Original cost
    - Estimated useful life
- Modified Approach
    - Complete, up-to-date inventory
    - Original cost
    - Regular condition assessment
    - Estimated annual cost to preserve assets
    - Documentation that target condition is maintained

Integrated Asset Management Systems

Integrating multiple sources of data maximizes efficiency and control and leads to smart decisions:

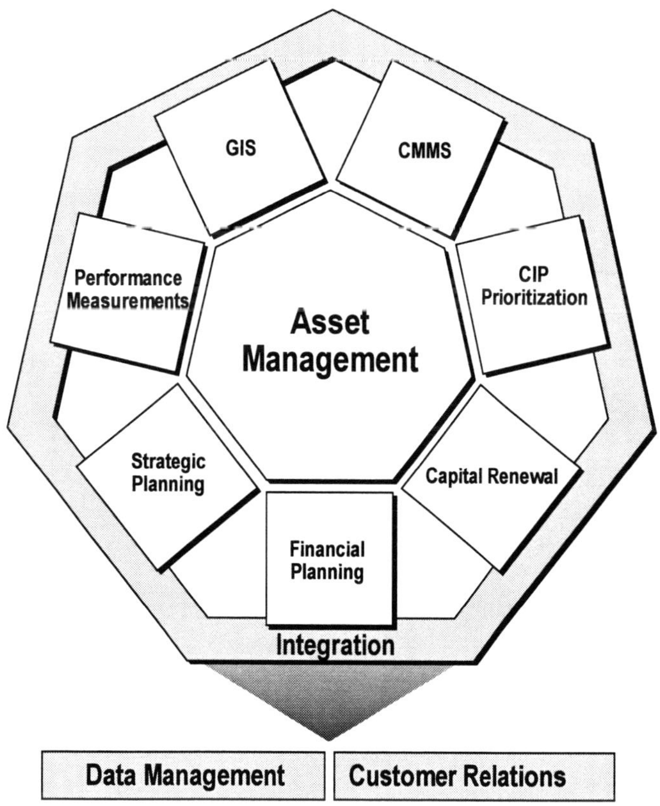

Meeting the Challenge and Making the Case

"Engage your stakeholders at all levels. The community is your best ally if you've got a problem. In bad asset management scenarios, it's harder to go to your community for support." -Workshop Participant

Organizational improvements and stakeholder communication enhance the success of asset management and improve a utility's competitiveness:. These steps include:

- Redesign business processes to advance asset management
- Develop the support of governing boards for necessary investments
- Develop stakeholder support for necessary investments
- Manage staff morale through difficult change to remain focused on the utility's overall goals

Asset management must be a key business process of every utility:

- Leadership is the most critical element to successful implementation and sustainability
- Eliminate "stovepipes" of various departments and align processes to serve a common mission

Chapter 3: Level of Tool Development

Review of Current Asset Management Tools

The workshop agenda included a presentation on asset management tools that are available today, as they relate to the modules in the asset management triangle diagram. The workshop facilitation team rated the tools using the evaluation criteria listed below, and scored each tool set using a "+/-" system, where "+++" was the highest possible score, and "---" the lowest possible score.

Tool Evaluation Criteria

- **Availability:** Numerous tools are available that achieve the same purpose
- **Accessibility:** The available tools are nonproprietary and easily obtained
- **Usefulness:** The tools are applicable to a wide range of conditions for the asset management activity
- **Acceptability:** The tool is in widespread use and is considered a "standard" in the industry
- **Simplicity:** The tool is easy to use; little if any customization is required
- **Cost:** Resources are likely to be available for the tool's implementation and ongoing use
- **Versatility:** The tool is applicable to more than one asset management activity
- **Documentation:** Procedures for using the tool are clearly detailed

The level of development of tools to support asset management was rated by the project team. Although many good tools are available, in none of the areas was the rating the highest--three pluses, pointing up the need to continue to develop good tools for asset management.

Strategic Asset Management Tools	*Tool Rating*
♦ Strategic planning	+
♦ Performance measures	++
♦ Level of service	+

Inventory and Condition Assessment	***Tool Rating***
♦ Information audits	+
♦ Setting up inventories	+
♦ Condition assessment methods	-
♦ Data collection	+
♦ Tracking/evaluating asset performance	+
♦ Quantifying condition and performance	+
♦ Modeling condition vs. performance	---
♦ Storing/displaying Geographic Information System	++

Maintenance Management	***Tool Rating***
♦ Integrating O&M	--
♦ Optimizing O&M programs	---
♦ Tracking O&M costs by asset	+
♦ Documenting preventive and predictive maintenance	-

Preservation, Renewal, and Replacement	***Tool Rating***
♦ Predicting future condition and useful life	---
♦ Quantifying risk levels	--
♦ Assessing risk vs. asset condition	--
♦ Evaluating alternative programs	---
♦ Calculating life-cycle costs	-

Some very good tools for asset management exist, although they may not yet be in widespread use.

Capital Planning and Financing	***Tool Rating***
♦ Projecting capital and operating costs	-
♦ Calculating revenue needs and rates	-

Financial Reporting Under GASB	***Tool Rating***
♦ Reporting to standards	-

Integrated Management Information Systems	***Tool Rating***
♦ Database integration	---
♦ Database management	--

Business Process Redesign and Stakeholder Outreach	***Tool Rating***
♦ Decision-making techniques	+
♦ Procedural manuals	++

Overall Perspective on Tools	***Tool Rating***
♦ Availability	+
♦ Accessibility	-
♦ Usefulness	+
♦ Acceptability	-
♦ Simplicity	-
♦ Cost	-
♦ Versatility	+
♦ Documentation	+

Chapter 4: Infrastructure Security

Special Report: Infrastructure Security

Like infrastructure owners everywhere, WERF sponsors, the EPA, and workshop participants are especially concerned with both accidental and intentional threats to the security of their assets. Securing infrastructure assets is a part of asset management; accordingly, workshop participants devoted a segment of the proceedings to security issues.

What might a terrorist threat to a wastewater system entail? Operators are concerned about on-site storage of chlorine gas; discharges into the drinking water supply; and the fact that the sewer system could provide access to key community locations. In addition, the sewer collection system might include other critical infrastructure in its conduits, such as telephone lines.

The workshop project team presented an overview of the current state of security in the water and wastewater industry as well as initiatives by the EPA and various associations. Highlights of this discussion are presented below.

Current State of Security

- Security threats from terrorism and related events are relatively new, so industry standard protocol is in its infancy
- Serious security practices are more evident in other locations, such as high-risk government buildings and nuclear power plants
- A utility's physical assets are typically dispersed, so standard approaches to security developed for enterprises with highly centralized assets are not likely to be a perfect fit for utilities

Security Initiatives

- AMSA, supported by an EPA grant, has developed two security checklists:
  - *Asset Based Vulnerability Checklist for Wastewater Utilities*
  - *Legal Issues in a Time of Crisis Checklist*
- AMSA is developing software to allow wastewater utility managers to assess their systems' vulnerabilities, evaluate risks, and identify countermeasures:
  - *Vulnerability Self-Assessment Tool (VSAT)*
- Water Environment Federation (WEF) is training wastewater professionals throughout the country in the asset-based approach to security management

"We link terrorist threat to reliability rather than asset management. We'd consider it in the same category as a threat from seismic activity."-Workshop Participant

Approach to Utility Security

- Identify assets
- Identify threats
- Determine criticality
- Identify countermeasures
- Assign vulnerability
- Determine P (loss)
- Determine risk level
- Define acceptability of risk
- Identify upgrades
- Perform benefit/cost analysis
- Prepare business continuity plan

Additional EPA-Funded Activities

- American Water Works Association (AWWA) training on general vulnerability assessment and countermeasures
- Direct grants to largest 360 water utilities for vulnerability assessments and planning
- Grants to states for training and monitoring progress
- State Operator Training Centers for course materials
- National Rural Water Association (NRWA) efforts to provide hands-on technical assistance
- Working with FEMA on emergency preparedness guidance for biological, chemical, and physical attacks

Water Information Sharing and Analysis Center (ISAC)

- EPA-sponsored secure, Internet-based communications tool for utilities to communicate with counterterrorism and law enforcement experts
- Membership organization requiring governing board approval for admission
- Annual fee based on utility size
- Being developed by AMWA; operational by Fall 2002
- Three main purposes of ISAC:
    - Disseminate early warnings and alerts concerning threats against cyber and physical infrastructure
    - Allow drinking water and wastewater utilities to share information with each other
    - Provide an opportunity for utilities to have security incidents analyzed by counterterrorism experts
- ISAC will provide:
    - Warning of potential and imminent threats to utilities
    - Reports of incidents nationwide
    - Incident trends
    - Possible responses to threats and attacks
    - Research on infrastructure protection

RAM-W Methodology for Water Utilities

- Risk Assessment Methodology-Water
- Developed by EPA, Sandia Labs, and AWWARF
- Performance-based approach evaluates effectiveness
- Prioritizes assets based on "detect-delay-respond" method
- "Design Basis Threat" under development
- 168 large utilities trained to date
- "Train the Trainer" program underway

Preparing for Sabotage

- Characterize systems: personnel, physical facilities, and customers
- Define threats and the likelihood of their occurrence
- Prioritize assets based on criticality of loss and countermeasures already in place
- Determine and prioritize vulnerabilities
- Identify and analyze countermeasures to protect against sabotage and minimize identified risks

A vigilant community is a key factor in securing infrastructure. Water and wastewater utilities, public health, and crime prevention networks can take advantage of "neighborhood watch"-type programs as part of their security measures.

Security Issues to Consider

- Physical destruction
- Biological and chemical contamination
- Coordination with public health officials
- Computer networks
- Deter-delay-detect
- Response capability
- Hazardous materials handling
- Insurance
- Vulnerability assessment redundancy
- New facility design
- Sewer collection system access
- Wastewater treatment plant
- Neighborhood watch
- Risk communication
- Vendors and contractors
- Hoaxes and psychological threats

The project team also presented a summary of EPA's recent announcement of $90 million in grants for water security:

- $53 million in direct grants to systems serving more than 100,000 people, supplying nearly half of all Americans
- $125,000 to $150,000 given for each of about 400 systems
- $4 million to wastewater systems for tool development, training for vulnerability assessment and emergency operations, technical assistance, investigation, treatment, and detection tools

Research Projects in Security

Although WERF intends to approach research in security separately from asset management, workshop participants were asked to suggest projects that address security. These were not evaluated for priority ranking by the workshop project team, but are presented here for review.

- **Protection of collection systems from unauthorized access:** Develop or identify methods and tools for preventing and detecting unauthorized access to wastewater collection systems, including illegal dumping, vandalism, or access for terrorist activities.

- **Strategies for building community partnerships to support infrastructure asset protection:** Many wastewater systems involve the public in capital infrastructure planning and siting decisions, but most have not established strong links with public health and law enforcement communities. These partnerships would strengthen a utility's ability to monitor threats and more effectively respond to emergencies.

- **Incorporating terrorist threats into asset risk assessment methodologies and emergency response plans for wastewater systems:** The definition of "risk" in the post-9/11 world has changed dramatically. Utilities should include security concerns in asset management planning, prioritization of critical assets, consequence planning, and implementation of asset management strategies.

Chapter 5: Pre-Workshop Survey Results

Pre-Workshop Survey

Prior to the workshop, the facilitator solicited responses to six questions from workshop participants. Respondents were asked to rank several issues based on an array of characterizations (i.e., higher, lower, difficult, not difficult, adequate, not adequate). The survey's purpose was to gauge participants' priorities and tailor the workshop programs to address them. A blank survey is provided for reference in Appendix E.

Overall, respondents indicated that implementing an asset management program is about equal in priority to other resource applications. There was little variation in the perspective on issues for Question 1.

Issues identified as higher-priority than asset management include public and worker health/safety, rapid emergency response, and existing or new regulatory standards. Changing work habits was identified as an additional high priority issue. Mediating various department needs was identified as much lower-priority than asset management.

Question 1: Which of the following agency objectives have higher or lower priorities for resources (management and staff time and agency budget) than implementing an expanded asset management program?

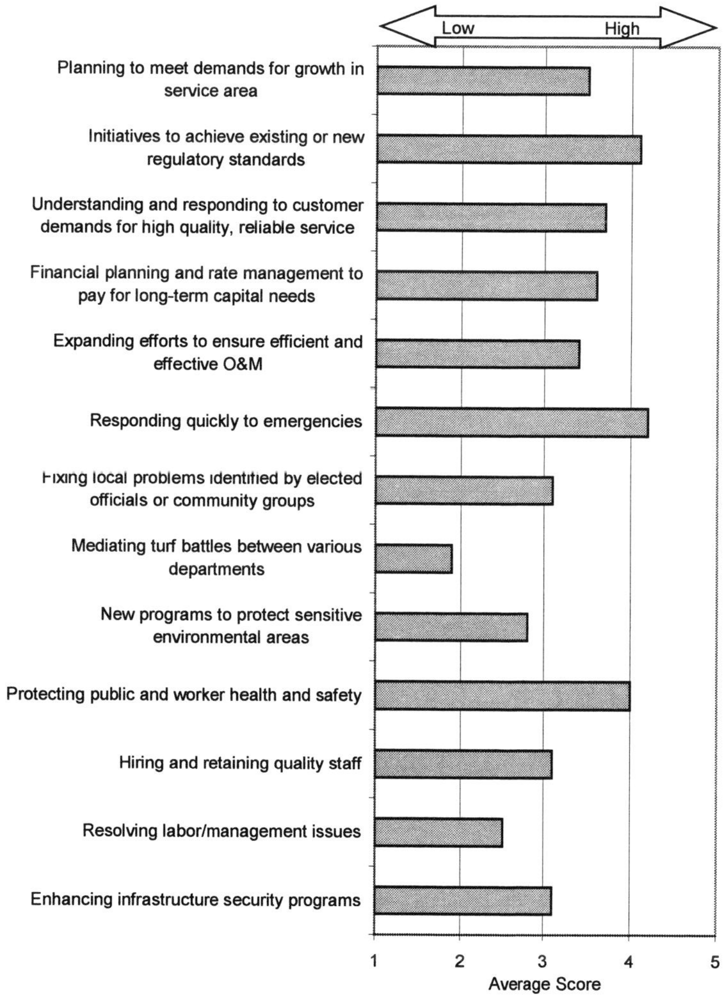

1=Much lower priority, 2=Lower priority, 3=About the same, 4=Higher priority, 5=Much higher

Responses for Question 2 were mixed. Respondents mostly indicated that the identified issues are significant in terms of a "driver" for implementing a more extensive asset management program. Several issues, such as GASB 34, stakeholder pressures, and economic development, were not viewed as key issues. Other driving forces include controlling capital costs and improving efficiency of the agency.

Question 2: What issues are driving your agency to implement a more extensive asset management program?

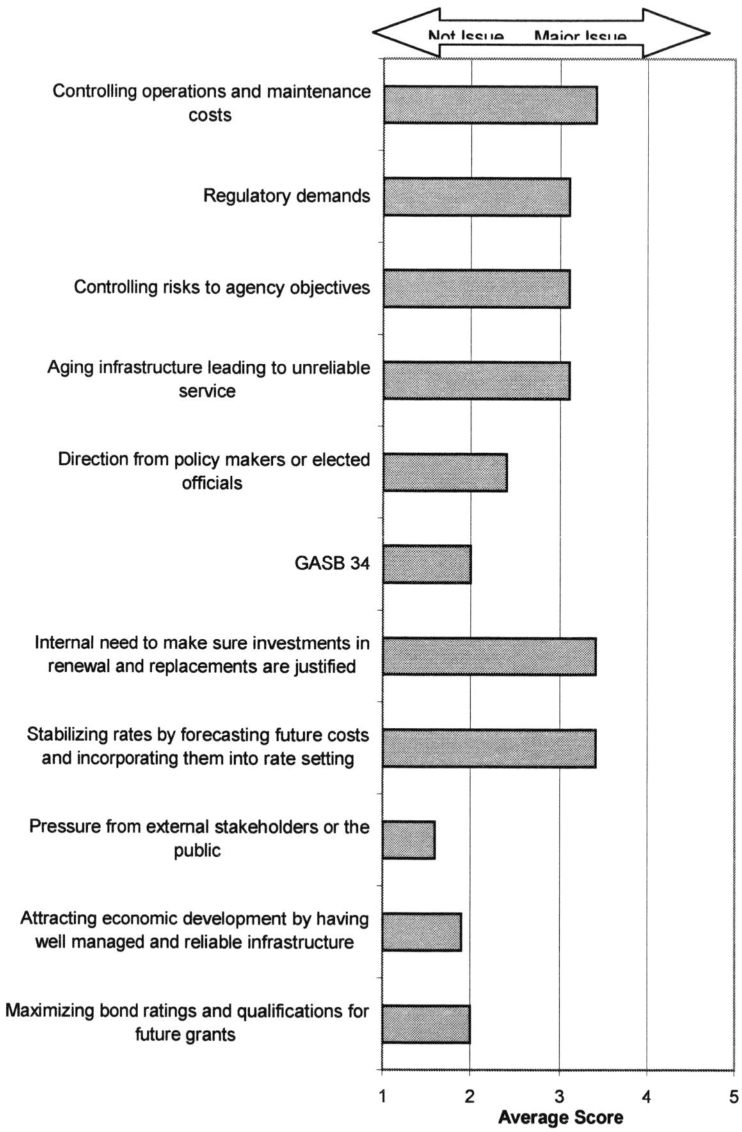

1=Not an issue, 2=An issue, 3=Significant issue, 4=Very significant issue, 5=Major driving force

Overall, respondents indicated that implementing the identified asset management process is "not difficult."

Respondents indicated that developing an asset management system that meets the needs of all users is somewhat difficult.

Question 3: How difficult do you believe it will be for your agency to implement the following asset management processes, considering the historical practices, organizational structure, executive/political and staff/management understanding of the issues, and internal culture of your agency?

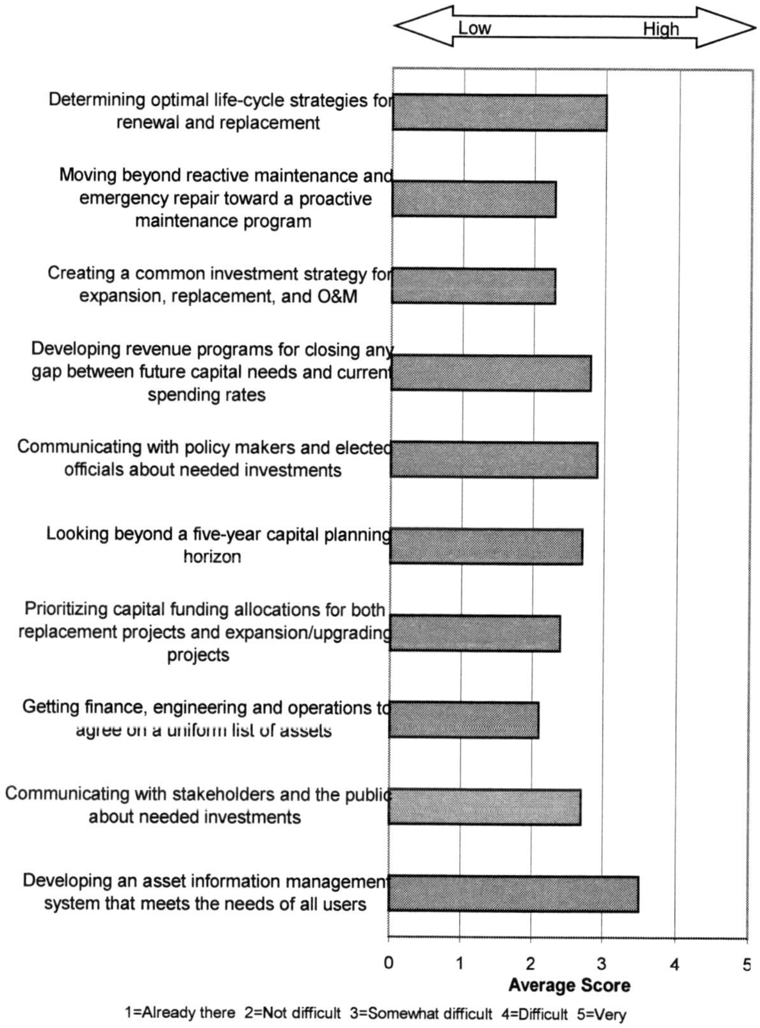

Overall, respondents seemed unconvinced that the necessary tools and technology could be obtained or used to implement the identified asset management processes. In particular, they had concerns about making life-cycle costing, renewal, and replacement, and making asset management understandable to internal and external constituents.

Identified tools needed include:

- *A means to compare utilities*
- *Proving amount of return on investment*
- *Standard asset decay curves*
- *Optimal mix of predictive maintenance (PdM)*

Question 4: Asset management requires technical and management techniques, analytical tools, decision-making processes, etc., in order to be successfully implemented. Based on the tools now employed by your agency and those that you are aware of, how confident are you that you will be able to obtain and use the necessary tools and technology to implement the following asset management processes?

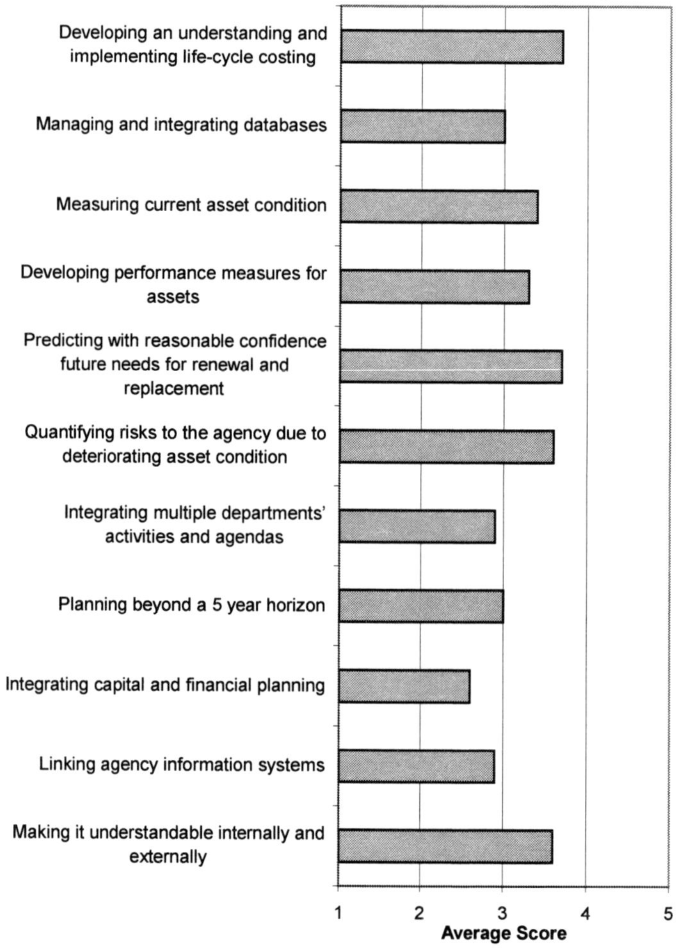

Already have tools , 2=Convinced , 3=Not convinced, 4=Undecided , 5=Do not believe

Overall, respondents indicated that they have started to implement most of the identified building blocks of asset management.

Some indicated that planning or significant work is needed for condition assessments, life-cycle decision for rehabilitation and replacements (R&R), and long-term planning for R&R.

Communication with other utilities was identified as another area where additional improvement is needed.

Question 5: Based on the current approach in your agency to the basic building blocks of asset management, how satisfied are you that your program meets the needs of your agency?

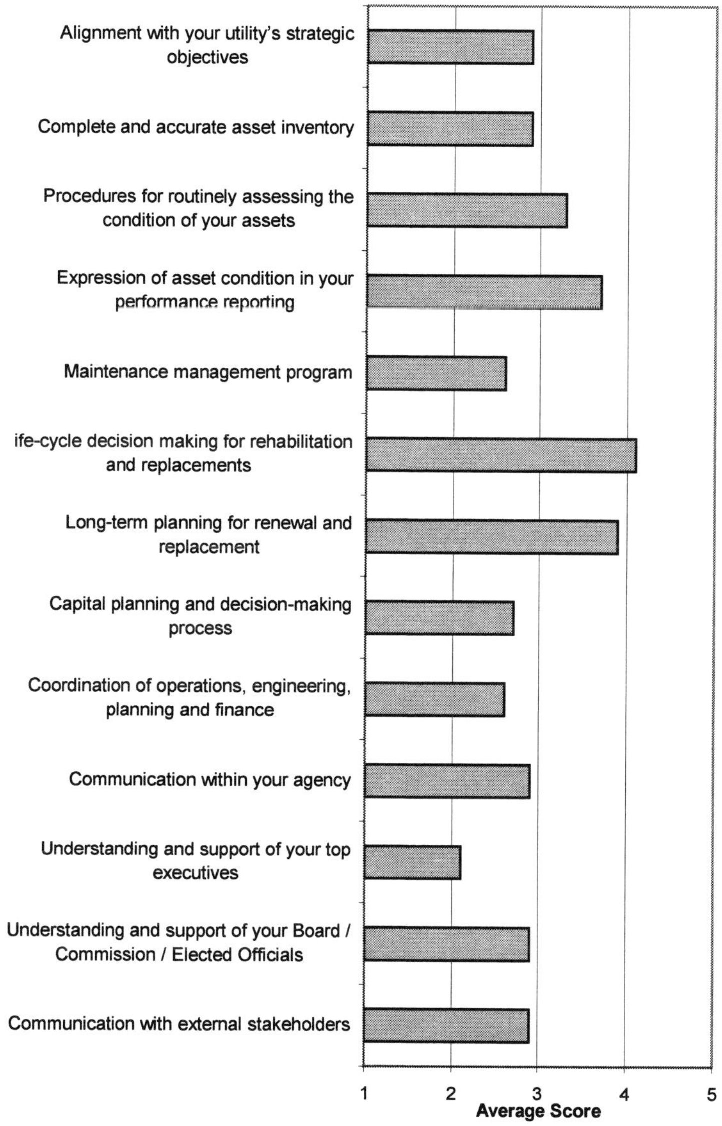

1=Already in place, 2=Almost in place, 3=Started to implement, 4=Just beginning to plan, 5=Needs significant work

Respondents indicated that improvement is needed for the majority of the identified security concerns, especially neutralizing or confining chemical or biological agents after release. Respondents indicated that systems for detecting intruders in facilities and communication systems for use before, during, and after an attack are largely satisfactory.

Question 6: The events of September 11, 2001 have forced utilities to reevaluate their assumptions about threats to water and wastewater systems. In addition to operational and staffing changes, most water and wastewater agencies are evaluating various technologies for preventing, detecting and minimizing impacts from intrusion and attack on their infrastructure. How confident are you that technology is available that is sufficiently effective and can be implemented at a reasonable cost?

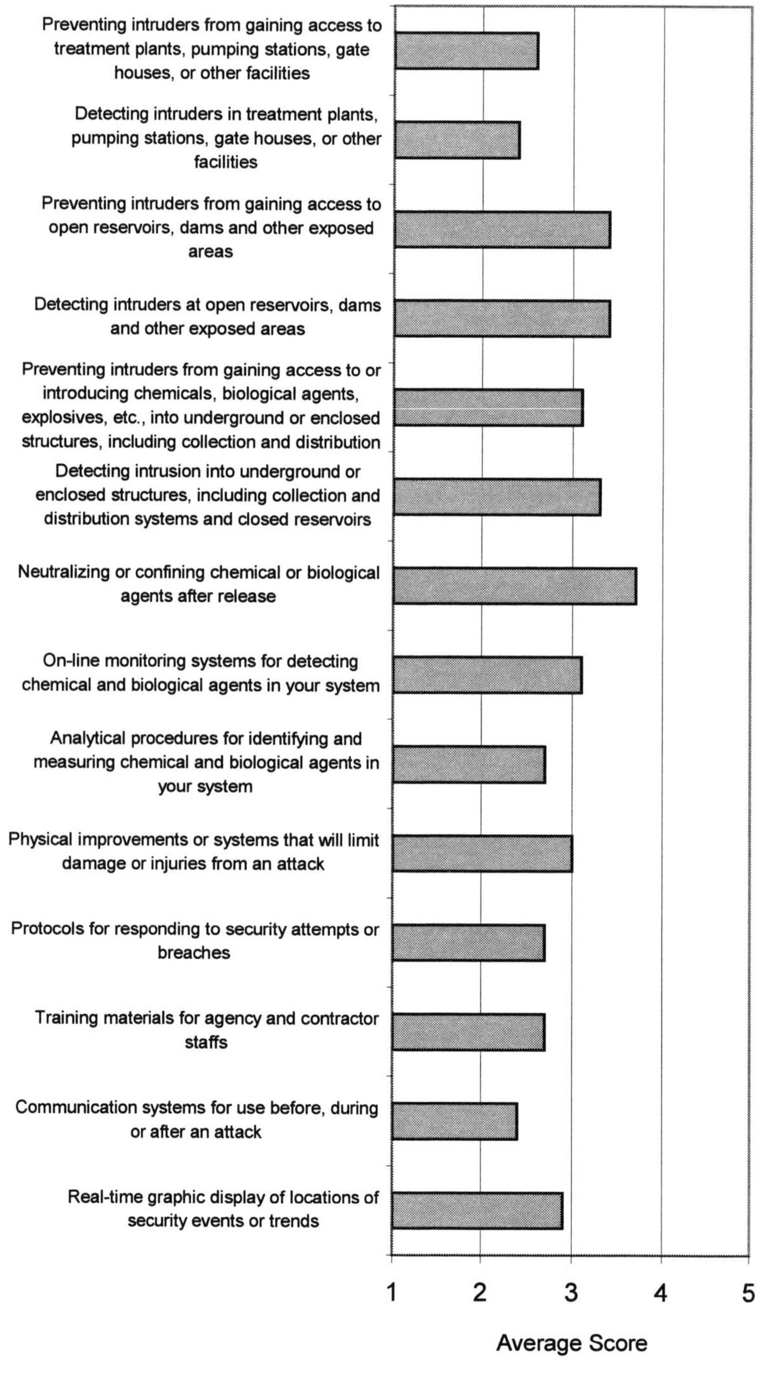

1=Satisfactory, 3=Improvement Needed, 5=Mostly inadequate or unaffordable

Chapter 6: International Participants' Presentations

It is well recognized in the United States that utilities in some other countries have applied asset management to their businesses with success. Compared to practices in these countries, asset management is less developed among U.S. water and wastewater utilities.

WERF invited utility representatives from Australia, Canada, New Zealand, and the United Kingdom to participate in the workshop and share their experiences with, and questions about, asset management. Highlights of their informal presentations are excerpted below.

Mick Bourke, Managing Director, City West Water Limited, Victoria, Australia

Business Model

In most areas of Australia, the major water and sewerage service providers are set up in a corporatized model. Under a national competition policy, they are self-funded and commercial in nature, run by skills-based boards of directors.

The focus in Australia is on customer service, and many utilities will establish and operate under a contract with their customers, in addition to their license targets.

The utilities operate under a license from the government that sets out the utility's objectives. Performance-based outcomes are measured against customer service, environmental issues, asset management, and financial targets.

Pricing for service generally is pay for use—with a fixed and variable, consumption-based component. For City West Water, the ratio is 20 percent fixed and 80 percent variable costs.

In the Australian business model, the state government is a shareholder in the utilities. The utilities return on the order of 65 percent of their before-tax profits to the government as a dividend. For example, we had an $80 million profit and returned a $50 million dividend. Debt-to-equity ratios are generally on the order of 30 to 50 percent. Local councils are not as large a force in water and sewerage as they are in other countries.

Regulatory Arrangements

Throughout Australia, utility regulation tends toward independent economic regulators. These are set up on a state-by-state basis. Again, they are strongly focused on customer service, asset

Everything is open: there are public reporting requirements regarding environmental discharges, water quality, infrastructure condition, and customer service performance.

In Melbourne, the regulatory structure is designed to drive comparative competition among the three water and sewerage retailers. A variety of public reports help customers compare the three retailers: the economic regulator, Essential Services Commission (ESC) Public Report; Water Services Association of Australia (WSAA) Facts publication; and WSAA Benchmarking and Drinking Water Quality Annual Reports. This has led to higher customer service levels than the minimum initially required. As a result, costs have decreased by 30 percent and service levels have significantly increased.

Asset management is a focus of economic regulators, who audit the utility businesses on a regular basis. The audits are deep, rigorous, and fairly intrusive. Regulators want to understand what water/sewer service is and see a program of improvement; they really fear catastrophic failure of infrastructure. Asset management programs make them more confident.

WSAA has benchmarked about a dozen companies in Australia to show where they stand vis-a-vis one another. It is expected this benchmarking will continue in the future.

management, environmental performance, and economic efficiency, i.e.; pricing is set to force efficiency and higher levels of service. However, there are few consistent linkages between economic, water quality, and environmental regulators, which gives rise to problems in coordinating standards for services in an economic-regulatory sense.

Most importantly, the regulatory climate leads to a high level of transparency to the community and a high level of community engagement in the utility.

Asset Management Drivers

There are four main drivers for asset management:

- Economic regulation of service delivery
- Customer service
- Capital scarcity and corporate risk (there is company and personal liability for poor management)
- Competition

Regulatory targets for customer service are established, including continuity of service and supply. The targets enumerate how many service interruptions will be tolerated in a yearlong period. In this way, customer service is a key driver of investments. Until a few years ago, 50 percent of our capital expenditure was focused on renewal of pipelines to meet continuity of supply demands.

The cost of capital and capital scarcity are other drivers for asset management. There is a mandate for efficient operations and cash returns to shareholders coupled with strong commercial tests before debt is allowed to be assumed. This makes the acquisition of funds for capital projects difficult. If a project cannot make the target rate of return then there has to be another compelling reason to finance it.

WSAA has studied utilities in Australia to examine their performance in customer service, civil maintenance, mechanical and electrical maintenance, etc. This benchmarking provides a way to share experiences and allows (and forces) utilities to examine how they are doing and how they can do better than the rest.

Approach to Asset Management

Asset management is the whole of the business process. The City West Water asset management program is driven and fully supported by its board of directors. The Asset Management Committee includes engineers, environmental specialists, finance specialists, the customer service manager, and the managing director. We work to ensure the whole staff has at least a rudimentary understanding of asset management and how it drives our business and operational decisions.

We approach asset management from an evidence-based, risk management, lowest life cycle cost perspective. Our program is not driven by asset age profiles; rather, it is based on asset condition, historical performance, trends, and judgments as to future performance.

Decisions are based on history, predicted future behavior, analysis of the drivers, and selection of the intervention method. This selection process would consider, for example, whether to renew a section of pipeline or wait and gather more evidence, or continue to operate and maintain the section.

Our risk and cost-benefit analysis includes the net present value of environmental impacts, service disruptions (and consequential damages), as well as operation and maintenance costs. We have developed a detailed but simple process to determine where to invest and when and how to intervene in the life of an asset.

We make decisions at a micro level, for example at the shutoff block level, rather than by asset class or by system. We have an array of detailed decision-making tools that help us do this: shutoff block assessment report (SOBAR), shutoff block assessment model (SOBAM), sewer condition model (SEWCON), distribution main assessment report (DMAR), meter management modeling, and sewer investment modeling.

We need to continue to collect and maintain even higher levels of information in order to continue to make evidence-based, risk-managed decisions.

These tools allow us to meet service standards, asset management audits, and financial return outcomes. As a result, we have better information on the infrastructure's condition and performance, allowing us to make more effective and efficient decisions on the intervention method. "Micro-managing" our system has reduced renewal spending by 80 percent and improved our customer service performance targets by 90 percent over the past three years.

Outlook: Five to Ten Years

- Customer opinion will be a prime driver in service standards. Customers must be involved in the service level cost/benefit tradeoff
- The industry approach to data gathering, data quality, and data analysis regarding condition assessment and predictive capabilities will evolve to a status of being truly valued by our businesses
- Decision tools will be well developed and widely used. More sophisticated decision tools will enable us to make better intervention decisions
- Scarce capital will be allocated to the genuine highest need
- Industry knowledge will be shared openly and will drive regulation. Cooperation among utilities will better harness the fruits of research and development

Sustainability will be better understood as a platform for the water and sewerage sector in asset management strategies

Generally, large municipalities have the ability to do asset management, while smaller municipalities are behind the curve. Overall, everyone needs more tools related to asset management program development and implementation.

Betty Matthews-Malone, Director, Infrastructure and Environmental Planning, City of Hamilton, Ontario, Canada

What's Happening at Various Levels of Government

In Canada, a national guide to sustainable municipal infrastructure is being developed at the federal level. This document will be similar to the national building and fire code. It will identify best program management practices for large and small municipalities relating to roads, water, sanitary sewer, and storm sewer infrastructure and address environmental management, decision making, and investment planning.

At the provincial level, recent deaths from an E. coli outbreak have spurred an examination of water quality regulations. There is not a Canadian version of GASB 34, but some bills have been introduced—e.g., Bill 135—that focus on sustainable water and sewer infrastructure management programs. Now that the hydroelectric industry has been deregulated, it is possible that the water and wastewater industry could follow that model.

At the municipal level, public works and water/sewer operations are run in a number of ways: some are private/public partnerships, others are completely private, although most are municipal. In Hamilton, we try to run ours as a business and we contract out some operations.

Case Study: The City of Hamilton

In January 2001, the city and six area municipalities were amalgamated to form the new City of Hamilton, located on the southwest shore of Lake Ontario, with a population of 470,000. By 1997, the municipalities had shifted from primarily being builders of infrastructure to predominantly caretakers. As growth really is not an issue, the focus is on taking care of the systems we have.

Essentially, we are looking at assets like people and taking a "demographic" approach. We assigned shorter life spans (20 years) for things like pumps, longer life spans (100 years) for pipes. We found that there was a "baby boom" of infrastructure that was approaching replacement phase.

The regional government began an inventory and assessment of its assets in the ground. The results showed that Hamilton's asset inventory was challenged. Working from a replacement value perspective, the asset inventory was valued at Cdn$3 billion. The city knew it was not investing enough in the system to maintain it.

While we spent a lot of time and money on this inventory and collected important data, it did not show us what we needed to do to get where we wanted to go. We needed a new strategic plan to deal with this staggering number and perhaps revalue our inventory. So now we are doing a very high-level strategic assessment, which so far has worked well to help us present complex information in an easy-to-understand way.

We extrapolated the asset demography information into capital life-cycle profiles covering a 100-year period and estimated that Cdn$140 million/year would be necessary to sustain the system. This represented double our current investment, and implies that a significant rate increase will be needed. We are working with the city council to raise rates to get to the sustainable funding level and have begun implementing a series of incremental rate increases that will gradually get us where we need to be. As increasingly detailed condition assessment data begins to come in, we can refine rate increases (and perhaps lower them).

Under the new amalgamated city government, the city has responsibility for all right-of-way assets and is developing an integrated asset management strategy for the entire public works portfolio, including water, sewer, roads, and bridges.

In a program like this, diversity and skills of staff are very important. For example, the manager of our asset management program is an accountant.

In the future, we will probably be moving from focusing on asset life profiles to focusing on asset condition profile. Even so, we also may revisit our initial estimates of asset life cycles as we move forward. We are currently looking at pilot software that integrates condition assessment with financial planning.

Ron Chapman, Head of Strategic Development, Advantica Technology, Loughborough, U.K.

Deregulation Spurred New Approaches Featuring Risk-Based Management

In the United Kingdom, deregulation has resulted in the formation of many combined and/or regional utilities. This privatization process has been a huge success for the end customer: rates are down and service levels are up. Investments in assets are also up; for example, water industry investments are up from £6 billion to £30 billion in a similar 10-year period.

Unlike before, utilities are rewarded for taking well-judged risks. We must accept, quantify, and manage risks. This is a new model for water and wastewater utilities that is more familiar to the oil and gas industry. There, they tend to put in shorter-lived assets—this provides more flexibility to respond to technology advancements and/or demand changes with new or different asset investments than is possible with longer-lived components.

We need to maximize the service(s) (and service levels) assets provide—it is the end results that matter.

Prior to deregulation and privatization, in the public sector there was no reward for taking or managing risks and system-related tradeoffs. Now, the regulations enable risks to be taken, which lets utilities make innovations. Under the new regime, utility managers can proactively evaluate and trade-off risk-cost-performance.

Customer Communication Should Focus on Quality, Not Concrete and Pipes

When we talk about maximizing service and minimizing costs, we need to communicate differently and more clearly about these trade-offs to our customers than we have in the past. For example, a utility should describe and justify a rate increase as necessary to improve its water quality, and not describe the increase in terms of the new treatment plant or expansion that the rate increase will pay for. Customers care about improving water quality, and they understand the linkage between rates and service levels.

There are 2,000-year-old underground pipes in Bath, U.K., that work very well.

Thinking About Underground Assets in a Different Way: Condition Assessment Doesn't Necessarily Relate to Statistics

How can we know if underground assets will perform or fail? We doubt that there are statistics to accurately predict this—for cast iron older than 40 years the curves don't apply. Age doesn't necessarily relate to failure rates.

Is there a way to assess condition that does predict failure? Advantica[1] operates 180,000 miles of 15- to 40-year-old cast iron pipe. We use statistical models with large data sets to quantify the likelihood of pipe leakage or failure. When we look at failure risks for a pipe, we take geography into account. We also estimate ancillary or collateral effects; for example, if that leak could cause gas to flow into a building, then what would be the consequence—in other words, we take a risk-based approach. We have used this approach to prioritize pipe replacement and have achieved a 25 percent reduction in catastrophic leaks (which is purposefully greater that the targeted reduction in general leaks.)

Risk Management is an Integral Part of the Business

One way to look at asset management is to equate it to the human body: we exercise to keep our hearts strong, and we don't smoke. Our approach to infrastructure management is parallel.

There is a different consequence of failure and risk among water- and wastewater-related choices—we can't engineer away all the risk, so we need to manage it. Here is an example of bringing risk management into water/sewer: you cannot build a storm drain large enough to handle the biggest possible storm, so you use statistics and make a judgment about how big it should be based on the trade-offs you are willing to make.

[1] Advantica is the research and consultancy wing of a company that moves gas over 180,000 miles of pipeline. Advantica deals with shippers and marketers, not directly with the end-users.

Anthony Wilson, General Manager, Community Assets
New Plymouth District Council, New Zealand

Regulatory Climate: Customer Accountability is the Main Asset Management Driver

Six Key Ingredients to New Zealand Asset Management Plans:

- *Life-cycle approach*
- *Long-term financial plan*
- *Defined (minimum) levels of service and performance monitoring*
- *Sustainable use of resources*
- *Risk management*
- *Continuous improvement*

Utilities are run differently in New Zealand: they are almost totally unregulated. In the 1970s and 1980s there was a complete reform of state organizations: most were privatized and all government subsidies were removed. We went from 800 government organizations to 78 local authorities. Five are big, 15 are of moderate size, and the remaining are small. We still probably have too many—there are still more than 50 local authorities that serve fewer than 50,000 people.

In 1989, commercial accounting practices were put in place, as were labor market reforms. As a result, water and wastewater utilities now outsource most of their activities. We integrate asset management into contract agencies that are the providers. Now, only policy and asset management really remain in-house.

Despite the lack of a regulatory framework, two major pieces of legislation provide a framework for asset management: the 1990 Resource Management Act and the 1996 Government Reform Act. These drive utilities' planning and program management.

The local authorities' programs are totally driven by the customer: their demands and preferences determine standards and establish service levels—and consequent prices.

We currently have 1,500 water supply entities serving 3.5 million people. We are subject to voluntary drinking water standards and mandatory wastewater quality standards. While there is no official economic regulator, the Auditor General makes annual reports to Parliament on the status of the local authorities.

Accountability is the main driver in our business. Under this regulatory structure, the local authorities' programs are totally customer-driven: their demands and preferences determine standards and establish service levels—and consequent prices. Our main focus is the customer. In contrast to our Australian neighbors, our utility neither receives federal funding nor pays dividends to the federal government.[2]

Well-Developed Asset Management Strategies

The asset management approach has been in play in our utilities since the mid-1980s. In the 1990, the pipeline asset management system became a common solution for all local governments that choose to use it.

[2] The federal government does provide 50-percent federal match for roads. So, a roads asset management program and for that matter, a parks and recreation AM program, are probably next.

We use contractors to conduct the evaluations of our pipes and feel this helps provide more consistent scores under the National Pipe Grading System. Still, we have noted that condition is not perfectly correlated with life-cycle performance. And performance is now taking precedence over condition in our management strategies.

In 1995, a National Asset Management Steering group (NAMS) was convened, composed of the auditor general, valuer general, Local Governments of New Zealand (a lobbying organization), infrastructure engineers, and water/wastewater association. The steering group's mission was to provide standard guidance and documentation for the following:

- Infrastructure asset management manual
- Valuation guidelines
- Training for practitioners/operators and politicians
- National pipe grading system

Integrating Asset Management Into Long-Term Financial Planning

Asset management plans are a means of compliance with long-term financial plans. Asset management is a *de facto* sustainability regulator. In 1995, New Plymouth's first plan showed that we needed to double water rates. This has happened—we doubled the price of water over five years. We are now moving to manage at a lower component level by life-term.

In 1996, new legislation mandated asset management plans as part of long-term financial plans. Legislation focuses on the levels at which assets are maintained and any decline in service level annually. In these plans, we look at customer demands versus the levels of service being provided. We must recoup the loss of value through depreciation and fund it at best value, which for underground assets is optimized replacement costs.

Asset Management is the Way in Which We Do Business

We have lots of asset management plans—some have been around long enough to have had 10 to 15 generations of the plan, and they are now almost annual reports. Customer service, quality, systems—all operate with real-time data validation in the field. Asset management does work. Continuous improvement is now in increasing focus in asset management plans.

We are now doing asset management plans for all public works. We're engineering them to track performance to customers, maintenance, quality, and contractors—all evaluations derived from asset management.

There are some institutional relationships that can change under this type of system. For example, politicians are just about obsolete in the process now. They are left with nothing else to decide after determining the level of service. After that, we have total control over rate setting. However, we predict that this won't last long and that they will want to get more involved in rate setting based on service levels.

The Customer is King

Asset management is driven by customer service—all else is a part of that. There is a minimum level of service that might be driven by regulatory requirements, but individual communities may differ on what they do or do not pursue beyond that. We have integrated customer inquiries into the asset management system.

How do customers express the level of service they want?

Through the political process customers get to decide the level of service that they want. The utility determines what that level of service will cost. Special industries or specific customers may demand different levels of service; therefore, customer differentiation is a critical issue in New Zealand. This approach also requires cooperation among authorities.

We are beginning to understand more about customer preferences and how to derive this outside of the ballot box. For example, Australia is doing a study on customer preferences that is finding customers are not demanding the highest levels of service. They understand the tradeoffs. And they are considering water in the context of the other services that they need to buy. The study also is finding that there are no differences in customer desires across various socioeconomic levels.

Chapter 7: Workgroup Activities and Research Recommendations

Participants' Goals

Early in the workshop, participants were invited to share their reasons for attending, or what topic they hoped to learn more about as a result of attending. This process was intended to encourage participants to think about the state of the practice in asset management and research needs. This discussion formed a foundation for the participants' contributions within their workgroups. While the specific goals of the participants varied widely, they fell into the several general categories as presented below.

"We need a primer on how to define and implement asset management. There is information overload in some areas, and vacuums in others. We need a road map." -Workshop Participant

The Maryland Center for Environmental Training is studying how to make asset management concepts work in small systems with limited resources.

Strategic Asset Management

- A good definition of "asset management"
- What is the international research agenda?
- Develop and implement a strategic approach
- Service industry—considering asset management in terms of what customers want and do not want
- System that would support increasing water rates. Justification and priorities
- Total asset management approach. How to link data issues with economies
- Risk-based approach to asset management
- Total comprehensive evaluation of sewer collection system is now underway so that they can prepare an asset management plan. Which strategy to adopt?
- What is happening globally in asset management?
- System expansion and reliability are issues along with asset management. Gain a broad view
- Get U.S. organizations to have holistic approach with sustainable outcomes, studying the Australian and New Zealand models
- In what research activities should EPA be involved?
- With a $320 million upgrade underway, how to incorporate asset management into this upgrade?
- Can U.K. and U.S. participate in research?
- Grow understanding of business success for water and wastewater
- Stabilization of user rates
- How to get catch-up funding
- Customer satisfaction—customers fund the system
- A New Zealand consultant did equivalent manual to AMSA's, which would be a good resource
- Establishing linkage between sustainable cities and strategic outcome
- Sustainable asset management

Strategic Asset Management, continued

- Recognize differences in asset management around the world
- Understand changes that must be made
- How assets are undervalued
- If we drive asset management from a risk perspective: network can be in good condition but with no redundancy; risk-driven; and condition/performance-driven
- Capacity for staff to tell strategic story in understandable terms. Frame it based on risk analysis, condition, etc., in a way that legislators can understand

"Today's models that predict the future show a tidal wave expenditure. Is this real, or will research show that such a spending spike is not necessary?" -Workshop Participant

Asset Retention

- Data collection
- Buried assets: what tools to assess condition?
- New and innovative techniques for condition assessment—one organization is developing these
- Better tools to make the decision: the "rehabilitate or replace" issue
- Pipe cleaning technology is "Neanderthal"
- Need to give people in the field better technologies with which to clean pipes
- O&M is an issue in developing countries
- Avoid spending spikes
- Condition assessment tools and successful predictive maintenance
- Better quality models and data related to asset performance and condition
- How to strategize implementation for IT, condition assessment, etc.
- Define the real useful life of an asset. What factors determine this?
- Using asset management for effective benchmarking, especially non-financial benchmarking. Must have national or international standards?
- How to validate a prediction model?
- Distinction between legacy and new requirements

Buried infrastructure mortality varies based on regional conditions. Utility managers need to know how to develop a system that can be applied uniformly, yet accounts for the variability of life cycles by region.

Policy/GASB 34

- How to stay ahead of the regulatory curve to benefit our ratepayers
- Congress will mandate asset management. Who will decide what it is and who will implement it?
- Quick evolution of where we are now; not to have asset management imposed on us
- How to set a goal to reach the point where the industry decides, rather than elected officials imposing it on us
- EPA compliance: get information to people so that they know what they need to do in order to be in compliance

"The regulatory environment will change. How should it? Instead of regulatory issues driving asset management, asset management should drive regulatory issues." -Workshop Participant

Policy/GASB 34 continued

- Local Government Environmental Assistance network
- Lgean.org: Populate this site with more information (ICMA helps run it) so that cities and towns of all sizes can get the information they need
- Cities with combined sewer systems—how to factor cost of environmental compliance into overall infrastructure needs? Washington, DC. cost of compliance estimated at $1 billion
- Be a better partner with policy makers
- Economic package attached to asset management that you can sell to short-term policy makers
- Money/GASB: rating agencies will be comparing how agencies manage assets (Moody's, Standard & Poor's)—this will become a metric
- Just because you depreciate assets, you still have to discuss the asset condition. Finance people consider that they're in fine condition and just a depreciable asset
- Rating agencies take asset management into account if you bring it up, but soon they'll start asking to see it
- Regulation is a sign of failure. This group should be ahead of the need for regulation

Integrated Asset Management Systems

- Integrating financial and non-financial systems
- We have a lot of information and a GIS of our system, but no way to prioritize asset management tasks
- What are the advances and gaps?
- Integrate information from IT from various sources
- Integrated data management

Business Process Redesign

- We are just starting our asset management program. We are finding that the hurdles we face are not technical ones; rather, we are challenged by how to change a silo organizational structure and run a better business
- How to prioritize implementation, considering the gaps?
- If the asset managers aren't doing these things, who is?
- As a finance guy, how to integrate the various players and their roles?
- Looking at asset management as part of the business
- Service levels and costs: how do you decide?

"In addition to the financial and technical sides, I'm concerned about the social side of the asset management equation." -Workshop Participant

"Selling the message is important. In New Zealand, national training workshops have really helped our utilities embrace asset management." -Workshop Participant

Outreach and Reporting

- Accountability
- Short-term sight of elected officials versus long-term sight of utility managers
- How to convince the government that there will be a return on the investment
- Industry database from which to tell Congress where we are as a group in asset management

Workgroup Approach

In order to most effectively discuss and develop ideas for WERF research projects, workshop participants formed groups of eight to 12 members each. The conference facilitator divided the groups based on the individual participants' areas of expertise and interest, though each included representation from utility managers, consultants, governments, and overseas colleagues.

The workshop facilitators led the workgroups through the process of developing criteria for asset management research needs, identifying barriers, and assessing gaps in programs; tools; and techniques.

The goals of the workgroups were as follows:

- Discuss and confirm the plus or minus ratings assigned to the tools discussed in the "Level of Tool Development" presentation
- Identify existing highly developed tools
- Identify gaps between existing tools and asset management needs
- Articulate research needs

Projects can be educational, research into better technology, or other. WERF wants to fund a series of projects that will build upon one another and lead to a conclusion, impact, and solution.

WERF Research Project Criteria

WERF staff offered the following priorities for funding research projects for conference participants' consideration (not ranked in order):

- Work that will help improve service levels
- Work that will save WERF subscribers money
- Work that will yield practical, implementable results
- Work that other groups are not likely to undertake, but which, if undertaken by WERF, would advance the state of the science/industry
- Work whose results would have wide applicability

Research Recommendations

Research project abstracts included a project title, objectives and rationale, benefits, budget estimates, duration estimates, and preliminary priority rankings.

As a result of their discussions over the course of the two-day workshop, the workgroups developed abstracts for 28 potential research projects. (See Appendix D for a preliminary listing of research projects as identified by the groups). Workshop participants were then given five votes each to apply at their discretion to nominate their top priorities for WERF consideration among the 28 potential projects.

Following the workshop, the project team then reviewed the participants' recommendations in detail and grouped the nominated projects into four categories:

- Strategy
- Asset retention
- Tool integration
- Outreach/business process redesign

The 10 project abstracts that were not ranked for priority consideration by WERF were not recommended for the following reasons:

- *Low number of votes*
- *No obvious linkages to higher-priority projects*
- *Objective covered in a better way by another project*
- *Significantly premature in current asset management climate*
- *WERF criteria not met*

Within these categories, which parallel the asset management triangle model, projects were combined, where appropriate, to reflect a logical research sequence or address duplication or near-duplication of objectives in two or more proposed projects.

The project team next assigned the resulting 11 projects to a decision matrix, which included the following factors:

- Meets WERF criteria
- Number of votes received at the workshop
- Objective aligned with results of pre-workshop survey
- Outcome would provide early or immediate benefits to the practice/industry
- Could leverage existing research to benefit the research

The following table summarizes the recommended research projects in order of priority. (See Appendix C for related details on the eleven projects.) The project numbers that begin with "W" are referenced to the complete table of project abstracts included in the appendix of this report.

	Project No. and Title (Original Workshop Project #)	WERF Criteria	Work-shop Votes	Survey Results Alignment	Early Benefits	Leverages Existing Information / Research Tools	Estimated Budget (000s)	Project Duration
1	Develop protocols for assessing the condition and performance of water and wastewater infrastructure assets and develop predictive models for correlating asset condition and performance (W-16, W-17, W-20, W-26)	●	2 6 15 9 (part)	●	●	●	$200 $200 $250 $650	12 mos. 12 mos. 12 mos. 36 mos.
2	Construct predictive life-cycle models for water and wastewater infrastructure that project life cycle costs and risks (W-4, W-24, W-27, W-26)	●	12 4 14 9 (part)	●		●	$150 $250 $200 $600	
3	Construct a template for preparing infrastructure asset management plans (W-9)	●	3	●	●	●	$75	6 mos.
4	Prepare guidance for asset management strategic planning (W-3)	●	14	●	●	●	$200	18 mos.
5	Develop practical methodologies for calculating life-cycle costs for water and wastewater infrastructure assets (W-25)	●	10	●	●	●	$250	18 mos.
6	Document case studies of improved customer service and satisfaction and reduced costs as a result of implementing asset management (W-8)	●	5	●	●	●	$200	18 mos.
7	Identify O&M best practices by asset category, condition, and performance requirements (W-21)	●	8	●		●	$250	24 mos.
8	Define best practices for the integration of water and wastewater databases (W-18)	●	6	●	●	●	$250	18 mos.

Project No. and Title (Original Workshop Project #)	WERF Criteria	Work-shop Votes	Survey Results Alignm nt	Early Benefits	Leverages Existing Information / Research Tools	Estimated Budget (000s)	Project Duration
9 Establish methodologies for determining water and wastewater asset value, compiling asset inventories, and capturing and compiling asset attribute information (W-6, W-14, W-19)	●	4 5 1	●	●		$100 $100 $150 $350	12 mos. 12 mos. 12 mos. 36 mos.
10 Develop a framework for asset management accountability (W-12)	●	5	●	●		$150	18 mos.
11 Assess the feasibility of establishing an Asset Management Standards Board for the water and wastewater industry (W-15)	●	6				$50	12 mos.

Example Sequencing for Recommended Projects

The following schedule shows how the 11 recommended projects could be arranged to roughly balance the projected annual spending over 4 years. This sequencing also considers starting longer-duration projects early, getting deliverables early from phased projects, and starting work later on projects that depend on deliverables from earlier projects.

Project No.	Year 1	Year 2	Year 3	Year 4	Total Budget
1					$650,000
2					$600,000
3					$75,000
4					$200,000
5					$250,000
6					$200,000
7					$250,000
8					$250,000
9					$350,000
10					$150,000
11					$50,000
Total Budget	$725,000	$705,000	$940,000	$655,000	$3,025,000

Projects Recommended for Later Consideration

The project team is not recommending WERF consideration of the remaining 10 projects at this time. These fall into three groups, as described below.

Worthy Projects that are Premature

These projects should be held for future consideration. They should not be advanced until work is done under the recommended projects because they would build upon those results.

W-5: Incorporating asset management into financial models and rate forecasts

> This project depends on establishing clear performance measures that can be incorporated into financial models.

W-7: Asset Management Benchmarking

> This project, as proposed at the workshop, is to benchmark all metrics and practices used for asset management. A number of the recommended projects are more narrowly focused on specific metrics and practices that the workshop participants felt were of critical importance. These are projects 3, 4, 6, 7, and 8. After these specific areas are benchmarked, a more informed judgment can be made as to the need for benchmarking the remaining practices.

W-10: Develop an asset management certification course

> The project team feels it is premature to establish such a program until more comprehensive industry standards are established and asset management practices are more mature in the water and wastewater industry.

W-11: Implementation of outreach and education tools for elected officials and/or agency staffs on asset management concepts

> The team agrees that this would be an extremely useful product for utilities. The project team understands that there is an initiative under way by another organization to do this. WERF should monitor the progress on this and consider raising it to a higher priority if the current initiative does not proceed.

Possibly Worthy Projects, but Much Lower Priority for Funding

W-2: Create optimal value infrastructure through development and adaptation of a best practice governance, regulatory, and business model

This project implies consideration of very fundamental changes to the structure of the water and wastewater industry in the U.S., including ownership, relationships and roles of different levels of government, economic mandates and pricing structures. The project team does not believe that there is a need at this time for such radical restructuring of the industry to accomplish the benefits of asset management.

W-28: Database on failure data

The project team doubts that at this time the benefit of such an industry-wide database would exceed the difficulties and costs of establishing and maintaining it. No sponsoring organization is apparent and it is unclear if utilities would participate. The proposal received no votes at the workshop

Not Recommended

These projects are not recommended by the project team for consideration by WERF. These are either policy, legislative or regulatory initiatives or projects to advance the development of specific machinery. The team understands that WERF does not sponsor such projects.

W-1: Data and methods for national goal and date target for initiating implementation of asset management at publicly owned treatment works (POTWs) and water utilities

This project received no votes at the workshop.

W-13: Develop regulatory incentives to achieve and implement asset management by POTWs and water utilities

This project received one vote.

W-22: Develop criteria for advanced technologies for cleaning sewers

This project received no votes.

W-23: Develop criteria for advanced technologies for sewer cleaning

This project is virtually identical to W-22 and received no votes.

Appendix A
Workshop Agenda

Time	Topic	Activity	Facilitator/ Presenter
Thursday, March 21, 2002			
8:30-8:50	Opening Remarks	Overview of WERF and EPA expectations from the workshop **Arlington Room (3rd Level)**	Linda Blankenship, WERF Steve Allbee, EPA Tyler Richards, PSC Chair
8:50-9:30	Introduction	Participant introductions Discussion of participants' expectations Begin listing of major issues	Ed Wagner, PI John Rogers
9:30-10:15	Presentation on asset management state of practice	Plenary presentation on background research on the state of practice in asset management for water and wastewater utilities	Ed Wagner Alan Ispass
10:15-10:30	Break		
10:30-11:00	Review of international asset management practice and issues in water/wastewater sector	Plenary presentation by international representatives	Mick Bourke, City West Water, Australia Betty Matthews-Malone, Hamilton, Ontario, Canada Ron Chapman, Advantica, United Kingdom Tony Wilson, New Plymouth District Council, New Zealand
11:00-11:30	Presentation on level of tool development	Plenary presentation rating the level of development of available tools for asset management	Ed Wagner
11:30-12:30	Discussion of research goals, criteria and weights	Group discussion to develop criteria and weights for prioritizing research	Linda Blankenship
12:30-1:30	Working Lunch Presentation on security management state of practice	**Fairfax Room** Plenary presentation on background research on the state c practice in security management fo water and wastewater utilities	 Elisa Speranza
1:30-2:30	Review participant survey data	Group presentation and discussion of participants' and industry needs	Alan Ispass
2:30-2:45	Break		
2:45-3:45	Breakout groups	Small group (4) discussion to focus on gaps and research needs. Each group will assign a leader and reporter	Project team will facilitate groups
3:45-5:00	Report back to large group	Presentation by breakout groups of lists of potential research projects	John Rogers

Time	Topic	Activity	Facilitator/ Presenter
Friday, March 22, 2002			
8:30-9:00	Review	Review of Day 1	Elisa Speranza
9:00-9:30	Decision process	Obtain consensus on decision process and criteria	John Rogers
9:30-10:15	Initial list of research projects	Review initial list by category of potential research projects from breakout groups	John Rogers
10:15-10:30	Break		
10:30-12:00	Break out by four clusters of asset management categories	Small groups discuss and rate projects by criteria. Groups will complete a definition and budget cost for each new project. Each group will assign a leader and reporter	Project team will facilitate groups
12:00-1:00	Lunch Break		
1:00-2:30	Report back to large group	Presentation by breakout group and discussion by large group of project goals and scope	John Rogers and Elisa Speranza
2:30-2:45	Break		
2:45 –3:45	Use decision model to prioritize research projects	Group participation to rank order projects	John Rogers
3:45-4:45	Implementation strategy	Discuss strategy for implementation by WERF	Ed Wagner
4:45-5:00	Concluding remarks	Comments on WERF implementing workshop conclusions	Chuck Noss, WERF
5:00	Adjourn		

Appendix B
Workshop Attendees

David Binning
Director, Planning and Engineering
Fairfax County, Virginia, Water Authority

Mick Bourke
Managing Director
City West Water Limited
Australia

Cuthbert Braveboy
Director, Department of Sewer Services
D.C. Water and Sewer Authority

Terrance M. Brueck
Vice President
EMA, Inc.
St. Paul, Minnesota

Warwick Busch
Market Sector Director, I-Business Consulting
Meritec Ltd.
New Zealand

Paul Causey
General Manager/District Engineer
Delta Diablo Sanitation District
Antioch, California

Ron Chapman
Head of Strategic Development
Advantica Technology
United Kingdom

Ralph L. Charlton
Director of Fiscal Services
Alexandria, Virginia, Sanitation Authority

Roger J. Dolan
President
Dolan Environmental Services
Martinez, California

Wayne Green
Director, WES/WWW Quality Control and System Planning
Toronto, Canada

Art Hamid
Vice President
Montgomery Watson Harza
Atlanta

V. Kenneth Harlow
Vice President, Director of Management Services
Brown and Caldwell
Irvine, California

Stephen T. Hayashi
General Manager/District Engineer
Union Sanitary District
California

Alan P. Landis
MSD Comptroller
Metropolitan Sewer District of Greater Cincinnati

Betty Matthews-Malone
Director, Infrastructure and Environmental Planning
Hamilton, Ontario

Jim Patterson
Vice President, Environmental and Information Services
Columbus, Georgia, Water Works

Stewart Ramsey
Senior Vice President
UMS Group North America, inc.
Parsippany, New Jersey

Eric de Rooy
Process Leader, Water Operations, Asset Management Division
Sydney Water Corporation
Australia

Duncan Rose
Parsons
Monticello, Florida

Russell Shaw
Principal
UMS Group
Australia

Andrew Speers
Director, CSIRO Urban Water
Commonwealth Scientific and Industrial Research Organization
Australia
Douglas M. Stewart

Asset Management Program Manager
Orange County, California, Sanitation District

Charles Vanderlyn
Project Manager
U.S. Environmental Protection Agency

Robert C. Weaver
Attorney, Kelly & Weaver
Washington, D.C.

Ray Yep
Deputy Operating Officer
Santa Clara Valley, California, Water District

Kevin Young
Manager, Corporate Planning & Government Regulation
Hunter Water Corporation
Australia

Observers

Doug Abbot
Senior Operations Specialist
Maryland Center for Environmental Training

Barry Benroth
General Engineer
U.S. Environmental Protection Agency

Walter Brodtman
Environmental Engineer
U.S. Environmental Protection Agency

Timothy J. Dwyer
Environmental Engineer
U.S. Environmental Protection Agency

Bijilsma Luitzen
Project Manager
World Bank – Netherlands Water Partnership Program/Assets Management

Robert S. Morrison
Senior Consultant
Jason Consultant International
Washington, D.C.

Anthony N. Tafuri
Research program Coordinator

U.S. Environmental Protection Agency

Project Subcommittee

Steve Allbee
Project Director, Gap Analysis
U.S. Environmental Protection Agency

Scott Haskins
Deputy Director
Seattle Public Utilities

Daniel K. O'Brien
Manager, Capital Programs
Massachusetts Water Resources Authority

Tyler Richards, Project Subcommittee Chair
Director of Water Reclamation
Gwinett County, Georgia

Michael Sweeney, Ph.D., P.E.
Director of Research and Productivity Assessment
Louisville and Jefferson County, Kentucky, Metropolitan Sewer District

Project Investigators/Facilitators

Edward Wagner, Principal Investigator
CH2M HILL
Parsippany, New Jersey

Lisa Bacon
CH2M HILL
Herndon, Virginia

Cara Bobchek
Capstone Communications
Vienna, Virginia

Alan Ispass
PA Consulting Group
Washington, D.C.

John Rogers
CH2M HILL
Philadelphia, Pennsylvania

Elisa Speranza
CH2M HILL
Boston, Massachusetts

WERF Team

Linda Blankenship, Deputy Director
Development and Subscriber Services

Amit Pramanik
Research Program Director

Roy Ramani
Project Manager

Appendix C
Details of Recommended Projects

Project No. 1

Workshop Project Nos. W16, W17, W20, W26 (part)

Title: Develop protocols for assessing the condition and performance of water and wastewater infrastructure assets and develop predictive models for correlating asset condition and performance

Objectives:

Phase 1: Survey assessment practices among U.S. and international water and wastewater utilities and in other industries (oil, gas, chemical, electrical, railroads, mass transit, etc.). Determine the availability, approachability and transferability of assessment and performance methodologies. Develop appropriate measures and metrics for condition and for performance that reflect levels of customer service for water and wastewater utilities. Condition metrics will be usable by utilities for reporting requirements of the Modified Approach under GASB Statement 34. Phase 1 report will be a stand-alone deliverable released for use by the industry.

Phase 2: Develop condition and performance assessment protocols using the measures and metrics developed under Phase 1.

Phase 3: Establish relationships for various asset classes/components to determine performance based on condition (including time-based relationships), and build appropriate models to correlate condition and performance.

Rationale:

There are no standardized guidelines for conducting condition assessments. Performance of an asset is not always directly related to condition.

Benefits:

This project will assist utilities to focus the approach for developing condition and performance measures; will assist utilities choosing the Modified Approach under GASB 34; will lead to comparable data sets and permit benchmarking metrics; will provide a methodology road map for proceeding; will enhance credibility of results; will minimize data collection for condition assessment to collect only relevant data and eliminate condition assessment where no correlation exists.

Project Duration: 36 months

Phase 1: 12 months

Phase 2: 12 months

Phase 3: 12 months

Budget: $650,000

Phase 1: $200,000

Phase 2: $200,000

Phase 3: $250,000

Project No. 2

Workshop Project Nos. W4, W24, W27, W26 (part)

Title: Construct predictive life-cycle models for water and wastewater infrastructure that project life-cycle costs and risks

Objective:

Phase 1: Determine best practices with regard to predictive models for costs and risks to customer service delivery based on current asset condition and alternative future O&M practices and rehabilitation investments. Assess their applicability to the U.S. water industry

Phase 2: Develop usable models for water practitioners that link, for principal classes of water and wastewater infrastructure: Asset condition information (current and projected), asset performance information (current and projected), risk evaluation and risk management options for environmental, social and financial issues, and life-cycle costs for asset intervention options, e.g., O&M, renewals, replacement, etc.

Phase 3: Demonstrate the models by applying them to a representative set of water and wastewater assets and a range of management options

Rationale:

Typically assigned economic life of assets bears little relationship to actual performance of an asset, particularly with regard to pipelines. Thus predictions of future investment are inaccurate. Predictive models, of necessity, are based on analysis of factors affecting asset performance in situ, and can be used to develop better investment strategies and target vulnerable/critical assets. Limited tools are available for predicting future asset condition, risks, and performance.

Benefits:

This project will provide linkages for asset condition, asset performance, risk levels, and lowest life cycle costs in one consistent set of models that will provide a clear view of future risks versus investment alternatives. This will provide highly-defensible forward investment projections to aid future pricing/rating, and justification to stakeholders.

Ultimate benefits can include more cost-effective investment strategies, better customer service, ability to choose the most appropriate repair/rehab techniques, and avoidance of catastrophic events.

Project Duration: 48 months

Phase 1: 12 months

Phase 2: 24 months

Phase 3: 12 months

Budget: $600,000

Phase 1: $150,000

Phase 2: $250,000

Phase 3: $200,000

Project No. 3

Workshop Project No. W9

Title: Prepare a template for preparing infrastructure asset management plans

Objective:

Develop a best practice standard for infrastructure asset management plans in the form of guidelines. The guidelines will include steps for preparing plans, basic and advanced plan structure and components, preferred presentation and format, GASB 34 compliance linkages, information flows to and from the plan, and checklists for review and audit of essential plan components.

Rationale:

Many utilities have implemented pieces of a comprehensive asset management program but are unsure about all the necessary components and the path for implementation.

Benefits:

This project will provide an effective vehicle for communicating the overall Asset Management strategy both internally and externally. The template will promote quality of content, logic, completeness, and consistency. It will also promote earlier completion of strategic plans and provide a basis for internal and external review and audit by demystifying how plan components relate to one another.

Project Duration:

6 months

Budget:

$75,000

Project No. 4

Workshop Project No. W3

Title: Prepare guidance for asset management strategic planning

Objective:

Provide a guidance manual for local utilities on how to implement an asset management program. Based on information from best practice performers, case studies will be presented documenting critical success factors, common elements, and guiding principles. The key components of an asset management strategy will be shown together with guidance for implementing the strategy.

Rationale:

The appropriate approach to implementing a comprehensive asset management program is to first develop a strategy for implementation. Many utilities are tempted to begin implementation of program components without first gaining an understanding of how they should be integrated and the institutional issues and obstacles that need to be addressed.

Benefits:

This project will allow utilities to accelerate the implementation of asset management programs by gaining an understanding of the experience and strategies of best practice performers. The guidance will provide a focused road map for the industry and individual utilities, leading to effective implementation of asset management.

Project Duration:

18 months

Budget:

$200,000

Project No. 5

Workshop Project No. W25

Title: Develop practical methodologies for calculating life-cycle costs for water and wastewater infrastructure assets

Objective:

Develop guidance for calculating life-cycle costs for various classes of water and wastewater assets. Included will be a definition of life-cycle costing (LCC). Interfaces of LCC with financial, economic and social aspects of decision-making will be defined to provide a global municipal perspective relative to CIP, R&R and O&M. Examples will be provided of approaches (basic, advanced) relative to level of detail and required data including review of activity based costing. Document case studies of real-life experience (what worked, what did not work, gaps) will be included.

Rationale:

Although LCC is often discussed and represented as an appropriate and comprehensive component of asset management, there is little documentation of its exact meaning and how it should be calculated. This project will assess practices in similar industries and propose an approach and methodology appropriate for the water and wastewater industry.

Benefits:

This project will promote a comprehensive approach to decision making that will optimize capital and O&M investments.

Project Duration:

18 months

Budget:

$250,000

Project No. 6

Workshop Project No. W8

Title: Document case studies of improved customer service and satisfaction and reduced costs as a result of implementing asset management

Objective:

Document examples of water and wastewater utilities that have improved customer service and satisfaction and have reduced costs by implementing asset management programs. Determine which management activities resulted in improvement and discuss their applicability, as well as constraints, relative to typical U.S. utility practices.

Rationale:

Case studies can be used as models for determining methods for measuring customer satisfaction and defining factors affecting customer satisfaction, obtaining a better understanding of the impacts of asset management on costs and value to the customers, achieving better responses to customer expectations, and reducing lifetime costs to customers.

Benefits:

This project will allow utilities to compare their own performance levels with those of other organizations using asset management. This information can be used to develop utility-specific strategies for implementation of asset management and for improving customer service and customer satisfaction.

Project Duration:

18 months

Budget:

$200,000

Project No. 7

Workshop Project No. W21

Title: Identify O&M best practices by asset category, condition, and performance requirements

Objective:

Determine benchmark O&M strategies and practices for water and wastewater utilities. O&M strategies and practices will be cataloged by type of asset, asset condition, and required performance. Best-in-class strategies and practices will be determined on the basis of cost and effectiveness.

Rationale:

It is generally acknowledged that the most cost-effective O&M programs provide a balance between proactive maintenance, such as predictive and preventive activities, and reactive maintenance, where assets are repaired as they fail. Determining how to achieve this balance is not well established.

Benefits:

This project will allow utilities to benchmark their O&M practices with identified best practices. Adoption of best practices will minimize O&M costs and improve O&M effectiveness. Utilities can make improved O&M decisions by asset category, resulting in overall lower O&M costs and reduced rate of asset failure.

Project Duration:

24 months

Budget:

$250,000

Project No. 8

Workshop Project No. W18

Title: Define best practices for the integration of water and wastewater databases

Objective:

Develop standardized methodologies for integrating the various databases commonly employed by utilities (GIS, SCADA, CMMS, customer service, financial management, human resources, etc). Prepare schematics and matrices of networking options and applications in use by utilities that are advanced in this area. Evaluate satisfaction among current users concerning the performance of the integrated networks. Criteria for performance satisfaction will include user-friendliness, expandability, degree of customization needed, cost and resources needed for maintenance, etc.

Rationale:

Most utilities possess multiple databases of the same or similar data, using different database applications. Integrating databases, or constructing interfaces between applications that will allow routine exchanges of data, would have numerous benefits. These benefits need to be assessed compared to the costs to implement and maintain an integrated system.

Benefits:

Would support knowledge-based decision making related to asset management, encourage uniformity among utilities, aid utilities in developing workable solutions through use of technology transfer, reduce duplicate data sets and multiple data entry, and improve accuracy.

Project Duration:

18 months

Budget:

$250,000

Project No. 9

Workshop Project Nos. W6, W14, W19

Title: Establish methodologies for determining water and wastewater asset value, compiling asset inventories, and capturing and compiling asset attribute information

Objective:

Phase 1: Develop valuation and depreciation guidelines for the U.S. water industry. Establish infrastructure valuation principles and issues. Provide a methodology for consistent valuation for depreciation, investment, financial reporting, and insurance purposes. Provide case examples and guidelines on determining original cost and replacement values. The Phase 1 report will be a stand-alone deliverable released for use by the industry.

Phase 2: Develop guidelines for practitioners to determine what and how data needs to be collected, structured and numbered for compiling asset inventories. The Phase 2 report will be a stand-alone deliverable released for use by the industry.

Phase 3: Conduct an analysis of best practices for inventory data capture and conversion. The Phase 3 report will be a stand-alone deliverable released for use by the industry.

Rationale:

Phase 1: Determining reasonably accurate costs—both original and replacement—is critical for long-term financial planning.

Phase 2: Decision making requirements in asset management drive the data structure hierarchy, data set detail, and numbering requirements. Inappropriate data sets will constrain outputs and can require expensive recapture or reconfiguration.

Phase 3: Field data capture and office data conversion are very expensive. Innovative techniques utilizing a range of technologies (APS, aerial mapping, ground penetrating radar, IT, digital imaging, etc.) are being used internationally to both minimize cost and improve data quality. Statistically valid techniques can provide high-confidence data for an entire asset group based on very limited sampling. Good practices exist in other industries.

Benefits:

Phase 1: Uniform procedures and methods will lead to greater credibility and assist utilities in financial reporting.

Phase 2: This project will help ensure appropriate data sets are collected by utilities for effective asset management. This will reduce incidents of expensive recapture or reconfiguration of data sets and make asset management systems more effective.

Phase 3: Data capture/conversion is one of the most expensive phases of creating asset management systems. Any savings in this phase will accrue high material benefits.

Project Duration:

Phase 1: 12 months

Phase 2: 12 months

Phase 3: 12 months

Budget:

Phase 1: $100,000

Phase 2: $100,000

Phase 3: $150,000

Project No. 10

Workshop Project No. W12

Title: Develop a framework for asset management accountability

Objective:

Define a framework for utilities to adopt an Environmental Management System approach to asset management. The framework will include self-audit procedures and templates for reporting status and effectiveness of asset management programs.

Rationale:

One critical objective for utilities adopting an asset management program is to enhance confidence among boards and commissions, elected officials, regulatory agencies, financial institutions, and other stakeholders. By developing industry-wide standards for internal audit and external reporting, utilities can document their stewardship of infrastructure assets.

Benefits:

Encourage and assist utilities in implementing an EMS-type approach to asset management. Enhance internal accountability and external confidence in a utility's management.

Project Duration:

18 months

Budget:

$150,000

Project No. 11

Workshop Project No. W15

Title: Assess the feasibility of establishing an Asset Management Standards Board for the water and wastewater industry

Objective:

Determine the interest and support within the U.S. water and wastewater industry to establish an asset management standards board. Input will be obtained from utilities, elected officials at the state and local levels, financial institutions, and regulatory agencies. Alternative approaches, structures, and functions of the board will be analyzed. The benefits and disadvantages of each will be assessed. The cost of supporting the board will be estimated and possible sources of resources will be explored.

Rationale:

Currently there are no broadly-accepted industry standards for asset management in the water and wastewater industry. Publicly-funded utilities could potentially benefit greatly if standards are developed by an industry-sponsored institution, rather that having no standards or having standards set through legislation or regulation.

Benefits:

The potential benefits of a standards board include: avoiding duplication of effort between industry sectors; encouraging a holistic approach to setting service levels/performance level; providing a forum for considering water as a common resource, not fragmented systems; providing a common industry voice; enhancing industry cooperation and discouraging a parochial approach; setting boundaries; and creating a forum for resolving differences.

Project Duration:

12 months

Budget:

$50,000

Appendix D
Preliminary Listing of All Projects As Developed by Workgroups

Project #	Title	Objective/Rationale	Benefits	Budget	Project Duration	Breakout Group Priority Level	Votes
W-01	Data and methods for national goal and date target for initiating implementation of asset management at POTWs and water utilities	1. Gain acceptance of asset management as a policy and tool by owners, elected officials, management, and the public 2. Establish date target for achieving full readiness to implement with national monitoring 3. Move towards POTW certification of local management plans and implementation	1. Greater acceptance by POTWs with interest and expertise 2. Demonstrates capability to implement to congress and others 3. Advances completion of a stronger fiscal partnership with federal government	$400,000	1.5-2.5 years	1	0
W-02	Create optimal value infrastructure through development and adaptation of a best practice governance, regulatory, and business model	This project will initially review existing governance/regulatory/business structures and access their impact on driving outcomes through asset management. From this adapt/develop a best practice model for the U.S. context. Model to cover: 1. Commercial corporate institutions under public shareholders 2. Industry framework covering federal/state/local responsibilities 3. Customer service levels represented through performance regulation 4. Economic regulation, pay for use pricing, targets for internal efficiencies 5. Clear accountabilities for performance 6. Mandate for effective AM to optimize infrastructure value	The model wil be used to inform the industry of the scale of change necessary to approach best practice AM. Forms the basis for influencing the public and political process over the long-term to achieve institutional change. Provides a discussion basis/framework to facilitate a pathway forward. (Co-cperative project with U.S., Australian/N.Z., UK and others; cooperation with WEF/AWWA)	Review $100,000 Model development $100,000	12 months	Start now	2

Project #	Title	Objective/Rationale	Benefits	Budget	Project Duration	Breakout Group Priority Level	Votes
W-03	Develop guidelines for best practice asset management strategic planning	1. Allow leapfrogging of the process through understanding the experience and strategies of best practice performers 2. Provide guidance to local utilities on how to implement an asset management program Steps/Outline 1. Identify relevant sectors (gas, electricity, water) 2. Select best practice performers in asset management 3. Collect available information from best practice performers 4. Follow up with a small number of on-site visits 5. Write up case studies ◆ Critical success factors ◆ Common elements ◆ Guiding principles ◆ Key components of asset management strategy 6. Prepare guidance manual of successful asset management and implementation	1. Effective implementation of asset management 2. A focused game plan for the industry and individual utilities 3. Determine change management process 4. Better customer service and lowest costs 5. Better team and strong morale (with time)	$100,000 (includes expenses and airfares)	Urgent project 8 months duration		14

Project #	Title	Objective/Rationale	Benefits	Budget	Project Duration	Breakout Group Priority Level	Votes
W-04	Methodologies that support optimal long term forecasting for infrastructure integrity and service delivery sustainability	Consistent methodologies for forecasting long-term investment needs to meet asset management requirements for long-term financial planning that takes account of the integration of all asset management drivers (PTO). Methodology must link: ♦ Asset condition information, current and projected ♦ Asset performance information, current and projected ♦ Risk evaluation and risk management options for environmental, social, and financial issues ♦ Lowest life-cycle costs for asset intervention options, e.g., O&M, renewals, replacement etc., including, to the extents required, NPV analysis of internal and external issues	This requires linkages for asset condition; asset performance, risk management; and lowest life cycle costs in one methodology providing a clear view of future investment needs. I.e.: provides highly defendable forward investment projections to aid future pricing/rating, and the justification to stakeholders. (High level methodology can be tiered down to form decision tools)	$80,000 Best practice search; workshop of key practitioners; development of a methodology manual	3 months		12
W-05	Incorporating asset management into financial models and rate forecasts	We must apply available models to forecast rates and financial impacts for utility asset management programs. The objective is to develop methodologies to meet service levels expressed as customer references/value choices. The project will develop a framework for customer service level preferences, assessment of value, willingness to pay, etc. The financial model will need to incorporate sustainable life-cycle approach. This project needs to be coordinated with AWWARF.	Stakeholders, such as users, providers, and policymakers, would have a clear picture of the impact of service level decisions. As an example, at any level, a new set of criteria could be entered to determine the impact on rates. This process will result in a better understanding of what is to be delivered and the resulting cost.	$100,000	6 months	Medium	2

Project #	Title	Objective/Rationale	Benefits	Budget	Project Duration	Breakout Group Priority Level	Votes
W-06	Develop natural infrastructure asset valuation standards and methodologies	Develop valuation and depreciation guidelines for the U.S. water industry ◆ Set out infrastructure valuation principles and issues ◆ Provide guidance on use of condition-based preservation of assets for the modified approach under GASB 34 ◆ Provide methodology for consistent valuation for depreciation, investment, financial reporting, and insurance purposes ◆ Provide working examples and guidelines on briefing/selection of values	Essential to understand the methodologies and GASB requirements to ◆ Monitor asset service protection (increase/decline) ◆ Demonstrate compliance for modified approach ◆ Have consistent residual lines/unit rates for tariff setting and for calculations	$50,000	3 months	Commence within 12 months (could be done in tandem with template as a single project under one working party for cost savings)	4
W-07	Asset management benchmarking	Develop key metric and practice benchmarks for asset management. Establish common definitions and framework for reporting performance, asset valuation, condition, service levels, and other attributes. Incorporate existing benchmarks and performance indicators of international practice leaders. Provide common repository of information on assets for water/wastewater industry.	◆ Help drive focus and improvement in asset management in water/wastewater industry ◆ Provide basis for comparing utility performance and characteristics for asset management across utilities ◆ Support and complement "benchmarking clearinghouse" for water/wastewater	$150,000 2 project workshops, 30 participants (national, international) AWWA, WEF, AMSA, EPA and others may be potential funding sources	6 months		3

Project #	Title	Objective/Rationale	Benefits	Budget	Project Duration	Breakout Group Priority Level	Votes
W-08	Case studies of improved customer service and customer satisfaction as a result of asset management	This project will allow utilities to compare their own performance levels with those of other organizations using asset management. This information can be used to develop strategies for implementation of asset management and for improving customer service and customer satisfaction.	Case studies can be used as models for: ♦ Methods for measuring customer satisfaction and determining factors affecting customer satisfaction ♦ Better understanding impacts of asset management on costs and value to the customers ♦ Reducing lifetime costs to customers ♦ Better responses to customer expectations	$100,000 2 to 3 workshops involving international participants	9 to 12 months	Start immediately	5
W-09	Template for developing infrastructure asset management plans	Develop a best practice standard for infrastructure asset management plans in the form of guidelines including: ♦ Steps for preparation ♦ Basic and advanced structure and components ♦ Presentation and format issues ♦ GASB 34 compliance linkage ♦ Information flows to and from the plan ♦ Checklists for review and audit ♦ Extreme examples	Provides an effective vehicle for communicating overall AM strategy both internally and externally (in shortened form). The standard will promote quality of content, logic, completeness, and consistency. It will also ensure earlier completion of central plans and provide a basis for internal/external review and audit. Also demystifies how the components of AM fit together.	$30,000 Literature search and development (2 drafts) $10,000 Working party review meetings (2) $40,000 Printing & publication (recover from sales)	3 months (using existing Australian/UK material)	Top priority in order to (a) support immediate understanding for how the key AM components link and are effectively reported to stakeholders to support better decision making, and (b) identify gaps in current information and strategies for inclusion in the AM improvement programs under the plan.	3

Project #	Title	Objective/Rationale	Benefits	Budget	Project Duration	Breakout Group Priority Level	Votes
W-10	Develop an asset management certification course	Design, prepare, and deliver a training course that results in certification. a) Present a basic AM approach b) Identify and develop the basic skill sets needed for effective AM c) Present the skills in modular format, e.g., O&M, engineering, planning, finance, IT, customer service, products and services d) Determine the method of course delivery, including exchange progress, large utility mentoring, conferences, courses and Internet	1. Raise the baseline level of awareness and skills for AM 2. Promote proactive and predictive approaches to achieving desired service levels 3. Promote a teamwork approach for utilities or related departments by increasing the knowledge of AM and knowledge of essential supporting roles	$250,000	2 years	Comes after identification of best practices and benchmarks	3
W-11	Implementation of outreach and education tools for elected officials and/or agency staffs on asset management concepts	The development of education materials targeted at elected officials and/or agency staffs to understand the basics of asset management. This could include updating existing materials such as videos, fact sheets, and brochures (e.g., NZ video & outlines). The elected official materials should look to both agency policy as well as education of state and federal officials concerning needs and uses of AM. These materials would allow all users to understand the need to properly manage infrastructure resources to understand risks associated with AM decisions.	♦ Broader knowledge at all levels of an organization ♦ Better public decisions – internally and externally ♦ More cost effective operations ♦ Faster agency staff acceptance of AM – support cultural change ♦ Buy-in to management philosophies and strategies ♦ Organizational alignment ♦ Expansion of concepts and knowledge within the industry ♦ Development of consistent understanding of AM within water and waste industry	$250,000-$500,000 (one time)	6 months	High (Rollout ASAP)	2

Project #	Title	Objective/Rationale	Benefits	Budget	Project Duration	Breakout Group Priority Level	Votes
W-12	Internal and public accountability framework for asset management to assist decision makers	1) Accounting/Reporting – Develop cheap, simple POTW modification of asset management accounting focused on condition and performance 2) Verification/Audit – Develop audit or verification model 3) Oversight – Develop process models for elected officials and policymaker oversight, including public oversight	1) Relates accounting standards to POTW needs and asset management as practiced 2) Leads implementation of policymaker and public oversight GASB & GFOA	$500,000	2-3 years	1	5
W-13	Develop regulatory incentives to achieve and implement asset management by POTWs and water utilities	1) Identify regulatory impacts and opportunities 2) Prepare specific proposals to promote/ improve asset management such as a substitute for such EPA initiatives as CMOM in the SSO rule, and new or increased legal responsibility standard including for liability, such as due diligence in implementing asset management plan	1) Provides incentive for asset management 2) Internalizes CWA objectives into POTW management 3) Reduces regulatory costs on EPA and POTWs with accountability	$300,000	1 year	1	1
W-14	Guidelines for compiling asset inventories – structure, numbering, attributes	Decision making requirements in asset management drive the data structure hierarchy, data set, and numbering requirements. In appropriate data sets will constrain outputs and can require expensive recapture or reconfiguration. This project will provide guidelines for practitioners to determine what and how data needs to be collected, structured, and numbered to aid decision making.	This project will help ensure appropriate data sets are collected by utilities for effective asset management. This will reduce incidents of expensive recapture or reconfiguration of data sets and make asset management systems more effective.	$100,000-$150,000	1 year (2003)	High	5

Project #	Title	Objective/Rationale	Benefits	Budget	Project Duration	Breakout Group Priority Level	Votes
W-15	Need benefits and function of an asset management multidisciplinary standards board	1) Avoid duplication of effort between industry sectors 2) Apply consistent rank/rating needs across water/wastewater and storm water sectors 3) Holistic approach to setting service levels/performance level 4) Consider water as common resource not fragmented system	1) Creates common industry voice/co-operation 2) Overcomes parochial approach 3) Sets boundaries and forum for resolving differences 4) Establishes uniform benchmark metrics 5) Provides forum for setting priorities 6) Standardizes condition and performance assessment ranking systems	$500,000 -$1 million	18 months - 2 years	Immediate	6
W-16	Standardized condition and performance assessment	1) Focus the approach 2) Create comparable data sets/benchmarking metrics 3) Across industry sectors vs. separate (parochial application) 4) Methodology road map for proceeding (integrating) third party would be comfortable with data results 5) Starting point for condition improvement	1) Transparency for customers 2) Transparency for regulators	$1 million – $2 million	2 - 3 years	Immediate	2

Project #	Title	Objective/Rationale	Benefits	Budget	Project Duration	Breakout Group Priority Level	Votes
W-17	Develop correlations between/among asset conditions and performance/ serviceability	Establish relationship for various asset classes/components to forecast/predict performance base on condition (including time-based relationships), and build appropriate models.	1) Minimize data collection for condition assessment to collect only relevant data and eliminate condition assessment where no correlation exists 2) Predict performance (current and future) for relevant assets	\$500,000-\$750,000	Phase 1: 6-9 months (\$250,000) Phase 2: 12-24 months (\$250,000-\$500,000)	Phase 1: establish relationships (primary research/field data collection; secondary research/ identify existing best practices) Phase 2: Build models and verify with pilot studies	6
W-18	Define best practices for the integration of databases (GIS, SCADA, Fin. Mgmt., HR, etc) to enable knowledge-based decision making throughout the utility, including asset management	Develop schematics and matrices of networking options and applications in use w/utility evaluation of performance satisfaction. (user friendly, expandability, degree of customization, etc)	Would enable comprehensive knowledge-based decision making related to asset management ♦ Would encourage uniformity between utilities ♦ Would aid utilities in developing workable solutions through use of technology transfer	In-kind contribution from 10-12 leading utilities & \$100K for contractor	6-12 months		6
W-19	An analysis of best practice for inventory data capture and conversion	Field data capture and office data conversion are very expensive. Innovative techniques utilizing a range of technologies (APS, aerial mapping, ground penetrating radar, IT, digital imaging, etc.) are being used internationally to both minimize costs and improve data quality. Statistically valid techniques can provide high confidence data for an entire asset group based on very limited sampling. Good practices exist in other industries.	Data capture/conversion is one of the most expensive phases of creating asset management systems. Any savings in this phase will accrue high material benefits.	\$150,000 - \$200,000	Urgent – 18 month time-frame	Very high priority to match current status of U.S. AM practice	1

Project #	Title	Objective/Rationale	Benefits	Budget	Project Duration	Breakout Group Priority Level	Votes
W-20	Analyze existing technology for condition/performance assessment – determine gap (including identification of research where no relevant technology exists) technology transfer	Survey other industries (oil, gas, chemical) to determine the availability, approachability, and transferability of assessment/performance technology. Identify GAPS in technology available overrun, technology needed in water and wastewater. Develop technology to full GAP.	Transferability can jump start assessment. 1) Contribute to national standards 2) Allow earlier identification of needs 3) Justify capital needs	$175,000	18 months	1) Identification of existing tech – High 3-6 months 2) Identify GAPS – High 3) Define research on technology to fill GAP-Medium	15
W-21	Identify O&M best practices by asset category, condition, and performance	Identify O&M strategies and practices by type of asset, it's condition and required performance and determine best in class. Best in class determined on basis of cost and effectiveness to trade offs. Utilities can benchmark their O&M practices with identified best practices. Adoption of best practices will either minimize O&M costs and/or improve effectiveness, while extending the useful life of the asset.	Improved O&M decisions, by asset category, resulting in overall lower O&M costs and reduced rate of asset failure	$250,000	1 year	7	8
W-22	Develop criteria for advanced technologies for cleaning sewers	Determine shortfalls in current sewer cleaning technologies and the gaps that need to be filled in order to improve the effectiveness of current technology, while maintaining or reducing costs. Consideration to be given to equipment footprint, cleaning frequencies, potential damage to sewers, effective length of pipe that can be cleaned from a single location, worker safety, and types of blockages and debris removal.	Reduce social impact and cost of maintaining a sanitary sewer collection system. Reduce frequency of overflows, regulatory violations, and improve customer service level	$100,000	1 year	5	0

Project #	Title	Objective/Rationale	Benefits	Budget	Project Duration	Breakout Group Priority Level	Votes
W-23	Develop criteria for advanced technologies for sewer cleaning	The methods employed today have been in practice for the past 20 to 50 years. In some cases they can be destructive to the pipeline. In most cases the problem occurs after a short period of time. The equipment is usually large and noisy.	Cost savings, smaller equipment footprint, fewer overflows, reduced frequency (cleanings), less disruption to neighborhoods	$100,000	1 year	5	0
W-24	Develop methodologies for determining effectiveness of repair/rehabilitation on service life	The objective is to balance cost versus repair/replacement levels. We also want to determine if the renewal activity accomplishes the predicted extension in service life.	More effective use of funds, improved performance of asset, ability to choose the most appropriate repair/rehab technique	$100,000	1 year	8	4
W-25	Develop a practical application for life cycle costing (LCC)	♦ includes definition of what LCC is, can be, and offer s ♦ Examples of approaches (basic, advanced) relative to level of detail and required data including review of activity based costing ♦ Interface of LCC with financial, economic, and social aspects of decision making for a global municipal perspective relative to CIP, R&R and O&M ♦ Examples of real-life experience (what worked, what did not work, gaps)	Comprehensive approach to decision making	$250,000	1 year	9/10	10

Project #	Title	Objective/Rationale	Benefits	Budget	Project Duration	Breakout Group Priority Level	Votes
W-26	Develop risk-based lifecycle asset management strategies to meet defined service requirements	Propose appropriate service measures and metrics (cross-sectional survey of industry, customers, and governing bodies; regulatory research). Review methods of linking required service level with asset management strategies and practices in other industries (gas transmission and distribution, petroleum refining and pipelines, private water and electric utilities, railroads and mass transit). Formulate a model to link quantified service requirements with asset-based life-cycle risk management strategies.	Resources and spending levels tied directly to levels of service required by stakeholders	$500,000	3 years, start soon	10	9
W-27	Document predictive performance modeling work performed globally and assess applicability	Assigned economic life of assets bears little relationship to actual performance of an asset, particularly with regard to pipelines. Thus predictions of future investment are inaccurate. Predictive models, of necessity, are based on analysis of factors affecting asset performance in situ, and can be used to develop better investment strategies and target vulnerable/critical assets. The objective of this project is to determine best practices with regard to predictive models and to asses their implementability in the U.S.	More cost effective investment strategies, better customer service, avoidance of catastrophic events	$250,000	2 years	10 (high)	14
W-28	Database on failure data	Build up a picture of actual asset performance by creating a central database for recording, continuously, incidences of failure and causes of failure. Rationale is that this is a more rapid means of determining performance than developing predictive models.	Cost effective development of performance information	Set-up = $100,000 Management = $25,000/yr	1 year set-up, then ongoing	5	0

Appendix E
Sample Pre-Workshop Survey Form

Participant Name: ______________________________

Utility Name: ______________________________

This survey is designed to provide a profile of the current needs, drivers, challenges and obstacles that the water and wastewater agencies participating in the workshop face in implementing asset management programs and the agencies' perception of the adequacy of current practices and tools. If you are not employed by a water or wastewater agency, please give your perspective on these issues based on your experience and understanding about where most agencies are along the path toward a fully effective asset management program.

The results of the survey will be used at the March 21-22, 2002 workshop to provide a foundation for our discussions to identify needed research into new processes and tools that will assist and encourage agencies to implement asset management programs more quickly and efficiently. The results will be aggregated and no agency will be identified when the results are reported.

For each question, please enter the number corresponding to your response.

1. Priorities for agency resources	**5**	**4**	**3**	**2**	**1**
Which of the following agency objectives have a higher or lower priority for resources (management and staff time and agency budget) than implementing an expanded asset management program?	Much higher priority than asset management	Some-what higher priority than asset management	About the same as asset management	Lower priority than asset management	Much lower priority than asset management
Planning to meet demands for growth in service area					
Initiatives to achieve existing or new regulatory standards					
Understanding and responding to customer demands for high-quality, reliable service					
Financial planning and rate management to pay for long-term capital needs					
Expanding efforts to ensure efficient and effective O&M					

1. Priorities for agency resources	**5**	**4**	**3**	**2**	**1**
Responding quickly to emergencies					
Fixing local problems identified by elected officials or community groups					
Mediating turf battles between various departments					
New programs to protect sensitive environmental areas					
Protecting public and worker health and safety					
Hiring and retaining quality staff					
Resolving labor/management issues					
Enhancing infrastructure security programs					
Other (please explain)________________________					
Other (please explain)________________________					
Other (please explain)________________________					
Other (please explain)________________________					

2. Drivers for asset management	**5**	**4**	**3**	**2**	**1**
What issues are driving your agency to implement a more extensive asset management program?	Major driving force		Significant issue		Not an issue
Controlling operations and maintenance costs					
Regulatory demands					
Controlling risks to agency objectives					
Aging infrastructure leading to unreliable service					
Direction from policymakers or elected officials					
GASB 34					
Internal need to make sure investments in renewal and replacements are justified					

2. Drivers for asset management	**5**	**4**	**3**	**2**	**1**
Stabilizing rates by forecasting future costs and incorporating them into rate setting					
Pressure from external stakeholders or the public					
Attracting economic development by having well-managed and reliable infrastructure					
Maximizing bond ratings and qualifications for future grants					
Other (please explain)________________					
Other (please explain)________________					
Other (please explain)________________					
Other (please explain)________________					

3. Institutional challenges in managing assets	**5**	**4**	**3**	**2**	**1**
How difficult do you believe it will be for your agency to implement the following asset management processes, considering the historical practices, organizational structure, executive/political understanding, staff/ management understanding of the issues, and internal culture of your agency?	Very cifficult		Somewhat difficult		Already there
Determining optimal, life-cycle strategies for renewal and replacement					
Moving beyond reactive maintenance and emergency repair toward a proactive maintenance program					
Creating a common investment strategy for expansion, replacement, and O&M					
Developing revenue programs for closing any gap between future capital needs and current spending rates					
Communicating with policymakers and elected officials about needed investments					
Looking beyond a five-year capital planning horizon					

3. Institutional challenges in managing assets	**5**	**4**	**3**	**2**	**1**
Prioritizing capital funding allocations for both replacement projects and expansion/upgrading projects					
Getting finance, engineering, and operations to agree on a uniform list of assets					
Communicating with stakeholders and the public about needed investments					
Developing an asset information management system that meets the needs of all users					
Other (please explain)________________					
Other (please explain)________________					
Other (please explain)________________					
Other (please explain)________________					

4. Availability of necessary tools and technology for asset management	**5**	**4**	**3**	**2**	**1**
Asset management requires technical and management techniques, analytical tools, decision-making processes, etc., in order to be successfully implemented. Based on the tools now employed by your agency and those that you are aware of, how confident are you that you will be able to obtain and use the necessary tools and technology to implement the following asset management processes?	Do not believe sufficient and adequate tools exist		I'm aware of some tools but not convinced they are adequate		We already have the necessary tools
Developing an understanding and implementing life-cycle costing					
Managing and integrating databases					
Measuring current asset condition					
Developing performance measures for assets					
Predicting with reasonable confidence future needs for renewal and replacement					

4. Availability of necessary tools and technology for asset management	**5**	**4**	**3**	**2**	**1**
Quantifying risks to the agency due to deteriorating asset condition					
Integrating multiple departments' activities and agendas					
Planning beyond a five-year horizon					
Integrating capital and financial planning					
Linking agency information systems					
Making it understandable internally and externally					
Other (please explain)____________________					
Other (please explain)____________________					
Other (please explain)____________________					
Other (please explain)____________________					

5. Satisfaction with your asset management program	**5**	**4**	**3**	**2**	**1**
Based on the current approach in your agency to the basic building blocks of asset management, how satisfied are you that your program meets the needs of your agency?	Inadequate and needs significant work		We have started to work on it		These components are already in place, use the best available procedures and tools, and are fully integrated
Alignment with your utility's strategic objectives					
Complete and accurate asset inventory					
Procedures for routinely assessing the condition of your assets					
Expression of asset condition in your performance reporting					
Maintenance management program					

5. Satisfaction with your asset management program	**5**	**4**	**3**	**2**	**1**
Life-cycle decision making for rehabilitation and replacements					
Long-term planning for renewal and replacement					
Capital planning and decision-making process					
Coordination of operations, engineering, planning, and finance					
Communication within your agency					
Understanding and support of your top executives					
Understanding and support of your board/commission/ elected officials					
Communication with external stakeholders					
Other (please explain)________________					
Other (please explain)________________					
Other (please explain)________________					
Other (please explain)________________					

6. Security concerns	**5**	**4**	**3**	**2**	**1**
The events of September 11, 2001 have forced utilities to reevaluate their assumptions about threats to water and wastewater systems. In addition to operational and staffing changes, most water and wastewater agencies are evaluating various technologies for preventing, detecting and minimizing impacts from intrusion and attack on their infrastructure. How confident are you that technology is available that is sufficiently effective and can be implemented at a reasonable cost?	Available technology is mostly inadequate or unaffordable		Improvement is needed in effectiveness or cost		Available technology is satisfactory
Preventing intruders from gaining access to treatment plants, pumping stations, gate houses, or other facilities					
Detecting intruders in treatment plants, pumping stations, gate houses, or other facilities					

6. Security concerns	**5**	**4**	**3**	**2**	**1**
Preventing intruders from gaining access to open reservoirs, dams, and other exposed areas					
Detecting intruders at open reservoirs, dams, and other exposed areas					
Preventing intruders from gaining access to or introducing chemicals, biological agents, explosives, etc., into underground or enclosed structures, including collection and distribution systems and closed reservoirs					
Detecting intrusion into underground or enclosed structures, including collection and distribution systems and closed reservoirs					
Neutralizing or confining chemical or biological agents after release					
On-line monitoring systems for detecting chemical and biological agents in your system					
Analytical procedures for identifying and measuring chemical and biological agents in your system					
Physical improvements or systems that will limit damage or injuries from an attack					
Protocols for responding to security attempts or breaches					
Training materials for agency and contractor staffs					
Communication systems for use before, during, or after an attack					
Real-time graphic display of locations of security events or trends					
Other (please explain)________________					
Other (please explain)________________					
Other (please explain)________________					
Other (please explain)________________					

2002 WERF Subscribers

The following agencies and corporations are WERF's charter stormwater research supporters:

Stormwater Agency

California
Monterey, City of
Santa Rosa, City of

Colorado
Boulder, City of

Georgia
Griffin, City of

Kansas
Overland Park, City of

Kentucky
Louisville & Jefferson County Metropolitan Sewer District

North Carolina
Cary, Town of
Charlotte, City of/Mecklenburg County

Pennsylvania
Philadelphia, City of

Corporate

BaySaver Inc.
Vortechnics Inc.
WWETCO, LLC

State

Fresno Metropolitan Flood Control District, Calif.
Urban Drainage & Flood Control District, Colo.

Wastewater Utility

Alabama
Montgomery Water Works & Sanitary Sewer Board

Alaska
Anchorage Water & Wastewater Utility

Arizona
Gila Resources
Glendale, City of, Utilities Department
Mesa, City of
Phoenix Water Services Department
Pima County Wastewater Management

Arkansas
Little Rock Wastewater Utility

California
Central Contra Costa Sanitary District
Contra Costa Water District
Crestline Sanitation District
Delta Diablo Sanitation District
Dublin San Ramon Services District
East Bay Municipal Utility District
El Dorado Irrigation District
Escondido, City of
Fairfield-Suisun Sewer District
Irvine Ranch Water District
Las Virgenes Municipal Water District
Lodi, City of
Los Angeles, City of
Los Angeles County, Sanitation Districts of
Orange County Sanitation District
Palo Alto, City of
Riverside, City of
Sacramento Regional County Sanitation District
San Diego Metropolitan Wastewater Department, City of
San Francisco, City & County of
San Jose, City of
Santa Barbara, City of
Santa Rosa, City of
South Bayside System Authority
South Orange County Wastewater Authority
Union Sanitary District

Colorado
Boulder, City of
Colorado Springs, City of
Littleton/Englewood Water Pollution Control Plant
Metro Wastewater Reclamation District, Denver

Connecticut
New Haven, City of, WPCA

District of Columbia
District of Columbia Water & Sewer Authority

Florida
Broward, County of
Fort Lauderdale, City of
Gainesville Regional Utilities
JEA
Kissimmee, City of, Department of Water Resources
Miami-Dade Water & Sewer Authority
Orange County Utilities Department
Orlando, City of
Reedy Creek Improvement District
Seminole County Environmental Services
St. Petersburg, City of
Stuart Public Utilities
Tallahassee, City of
Tampa, City of
West Palm Beach, City of

Georgia
Atlanta Wastewater Services
Augusta, City of
Clayton County Water Authority
Cobb County Water System
Columbus Water Works
Fulton County
Gwinnett County Department of Public Utilities
Macon Water Authority
Savannah, City of

Hawaii
Honolulu, City and County of

Illinois
American Bottoms Wastewater Treatment Plant
Dupage, County of
Greater Peoria Sanitary District
Metropolitan Water Reclamation District of Greater Chicago
Wheaton Sanitary District

Indiana
Fort Wayne, City of
Indianapolis, City of

Iowa
Ames, City of
Des Moines Metro Wastewater Reclamation Authority

Kansas
Johnson County Unified Wastewater Districts
Unified Government of Wyandotte County/Kansas City, City of

Kentucky
Louisville & Jefferson County Metropolitan Sewer District

Louisiana
Sewerage & Water Board of New Orleans

Maine
Bangor, City of

Maryland
Anne Arundel County Bureau of Utility Operations
Howard County Department of Public Works
Washington Suburban Sanitary Commission

Massachusetts
Boston Water & Sewer Commission
Massachusetts Water Resources Authority
Upper Blackstone Water Pollution Abatement District

Michigan
Ann Arbor, City of
Detroit, City of

Grand Rapids, City of
Holland Board of Public Works
Lansing, City of
Owosso Mid-Shiawassee County WWTP
Saginaw, City of
Wayne County Department of Environment
Wyoming, City of

Missouri
Independence, City of
Kansas City Missouri Water Services Department
Little Blue Valley Sewer District
Metropolitan St. Louis Sewer District

Nebraska
Lincoln Wastewater System

Nevada
Henderson, City of

New Jersey
Passaic Valley Sewerage Commissioners

New York
New York City Department of Environmental Protection

North Carolina
Cary, Town of
Charlotte/Mecklenburg Utility Department
Durham, City of
Metropolitan Sewerage District of Buncombe County
Orange Water & Sewer Authority

Ohio
Akron, City of
Butler County Department of Environmental Services
Columbus, City of
Metropolitan Sewer District of Greater Cincinnati
Northeast Ohio Regional Sewer District

Oklahoma
Tulsa, City of

Oregon
Clean Water Services
Eugene/Springfield Water Pollution Control
Water Environment Services

Pennsylvania
Philadelphia, City of
University Area Joint Authority, State College

South Carolina
Charleston Commissioners of Public Works
Mount Pleasant Waterworks & Sewer Commission
Spartanburg Sanitary Sewer District

Tennessee
Cleveland, City of
Knoxville Utilities Board
Murfreesboro Water & Sewer Department
Nashville Metro Water Services

Texas
Amarillo, City of
Austin, City of
Dallas Water Utilities
Denton, City of
El Paso Water Utilities
Fort Worth, City of
Gulf Coast Waste Disposal Authority
Houston, City of
San Antonio Water System
Trinity River Authority

Utah
Salt Lake City Corporation

Virginia
Alexandria Sanitation Authority
Arlington, County of
Fairfax County Virginia
Hampton Roads Sanitation District
Henrico, County of
Hopewell Regional Wastewater Treatment Facility
Loudoun County Sanitation Authority
Prince William County Service Authority
Richmond, City of

Washington
Edmonds, City of
Everett, City of
King County Department of Natural Resources
Seattle Public Utilities

Wisconsin
Green Bay Metro Sewerage District
Madison Metropolitan Sewerage District
Milwaukee Metropolitan Sewerage District
Racine, City of
Sheboygan Regional Wastewater Treatment
Wausau Water Works

Australia
Sydney Water Corporation

Canada
Toronto, City of, Ontario

Mexico
Servicios de Agua y Drenaje de Monterrey, I.P.D.

United Kingdom
Yorkshire Water Services Limited

Corporate

The ADVENT Group Inc.
Alan Plummer & Associates
Alpine Technology Inc.
Aquateam–Norwegian Water Technology Centre A/S
Black & Veatch
Boyle Engineering Corporation
BPR CSO
Brown & Caldwell
Burns & McDonnell
The Cadmus Group
Camp Dresser & McKee Inc.
Carollo Engineers Inc.
Carpenter Environmental Associates Inc.
CDS Technologies Inc.
Chemtrac Systems Inc.
CH2M HILL
Clancy Environmental Consultants Inc.
Damon S. Williams Associates, LLC
Earth Tech Inc.
Ecolab Water Care Services
EMA Inc.
The Eshelman Company Inc.
Finkbeiner, Pettis, & Strout Inc.
Frontier Geosciences Inc.
ftn Associates Inc.
Gannett Fleming Inc.
GE Betz
Golder Associates Inc.
Greeley and Hansen LLC
The HACH Company
Hazen & Sawyer, P.C.
HDR Engineering Inc.
HNTB Corporation
HydroQual Inc.
Insituform Technologies Inc.
Institute for Environmental Technology & Industry, Korea
Jacobson Helgoth Consultants Inc.
Jason Consultants Inc.
Jordan, Jones, & Goulding Inc.
KCI Technologies Inc.
Kelly & Weaver, P.C.
Kennedy/Jenks Consultants
Komline Sanderson Engineering Corporation
Lawler, Matusky & Skelly Engineers, LLP
Limno-Tech Inc.
Lombardo Associates Inc.
Malcolm Pirnie Inc.
McKim & Creed
MWH
Odor & Corrosion Technology Consultants Inc. (OCTC)
ONDEO Degremont Inc.
PA Government Services Inc.
Parametrix Inc.
Parsons
Post, Buckley, Schuh & Jernigan
The RETEC Group
R.M. Towill Corporation
Royce Technologies
Sear-Brown
SYNAGRO
Tetra Tech Inc.
Trojan Technologies Inc.
URS Corporation
USFilter NATC
Wade-Trim Inc.
Weston Solutions Inc.
Woodard & Curran
Woodruff & Howe Environmental Engineering Inc. (WHEE)
WRc/D&B, LLC

Industry

American Electric Power
BP-Amoco Corporation
Chevron Research & Technology Company
The Coca-Cola Company
Dow Chemical Company
DuPont Company
Eastman Chemical Company
Eastman Kodak Company
Merck & Company Inc.
ONDEO Services
Procter & Gamble Company
Reliant Energy
Severn Trent Services Inc.
Shell Global Solutions
Thames Water Plc
United Water Services LLC

Board of Directors

Research Council

Stormwater Technical Advisory Committee

WERF Product Order Form

As a benefit of joining the Water Environment Research Foundation, subscribers are entitled to receive one complimentary copy of all final reports and other products. Additional copies are available at cost (usually $10). To order your complimentary copy of a report, please write "free" in the unit price column. WERF keeps track of all orders. If the charge differs from what is shown here, we will call to confirm the total before processing.

Name Title

Organization

Address

City State Zip Code Country

Phone Fax Email

Stock #	Product	Quantity	Unit Price	Total

Postage & Handling	
VA Residents Add 4.5% Sales Tax	
Canadian Residents Add 7% GST	
TOTAL	

Method of Payment:

(All orders must be prepaid.)

- ❑ Check or Money Order Enclosed
- ❑ Visa ❑ Mastercard ❑ American Express

Account No. Exp. Date

Signature

Shipping & Handling:

Amount of Order	United States	Canada & Mexico	All Others
Up to but not more than:	Add:	Add:	Add:
$20.00	$5.00*	$8.00	50% of amount
30.00	5.50	8.00	40% of amount
40.00	6.00	8.00	
50.00	6.50	14.00	
60.00	7.00	14.00	
80.00	8.00	14.00	
100.00	10.00	21.00	
150.00	12.50	28.00	
200.00	15.00	35.00	
More than $200.00	Add 20% of order	Add 20% of order	

*minimum amount for all orders

Note: Please make checks payable to the Water Environment Research Foundation.

To Order (Subscribers Only):

Log on to www.werf.org and click on the "Product Catalog."

Phone: (703) 684-2470
Fax: (703) 299-0742.

WERF
Attn: Subscriber Services
601 Wythe Street
Alexandria, VA 22314-1994

To Order (Non-Subscribers):

Non-subscribers may be able to order WERF publications either through WEF (www.wef.org) or IWAP (www.iwap.co.uk). Visit WERF's website at www.werf.org for details.